The Best of *Islands in the Clickstream*

by

Richard Thieme

Minneapolis MN

2021

Copyright © 2021 by Richard Thieme. All rights reserved. No portion of this book may be reproduced, printed, or distributed in any way without prior consent from the publisher.

Author photograph by Eli Omen. Copyright Eli Omen 2020, Creative Commons BY-NC-ND

Published by Exurban Press

Printed by The Workshop, Arcadia CA. Jeff Smith, Proprietor. Noel Smith, Layout/Production

ISBN 978-1-7362663-2-8

What Readers Say

"Richard Thieme is an extraordinary person in every sense. Only someone who has lived different lives and could cope with what those had in stock for him, who sees with soft eyes while reflecting everything around him, can express the meta layer of today's world and technology as the one thing that it is."
Felix Lindner (FX) of Phenoelit

"I quite liked your story. I've been reading science fiction lately and your story ("More Than a Dream," in *Mind Games*) was right up there."
John Updike, Author

"Richard Thieme discusses the unintended consequences of technological progress, as well as extrapolating the expected consequences. He not only discusses the new capabilities that will be available to humans, but also what impact living in the context of such capabilities will do to what it means to be human."
Brian Snow, Senior Technical Director, NSA (ret)

"Give me Richard Thieme. His mind is in orbit but his feet are on the ground."
Dan Geer, CISO, In-Q-Tel (CIA)

"Thieme is truly an oracle for the Matrix generation."
Kim Zetter, author of *Countdown to Zero Day*

"You are a practitioner of *wu wei*, the effort to choose the elegant appropriate contribution to each and every issue you address."
Hal McConnell, NSA

"When you read Richard Thieme, you believe in the matrix."
Sol Tzvi, Former Senior Security & Privacy Architect, Trustworthy Computing, Microsoft Israel

"Thieme's writing represents a glimpse into the inner workings of a most extraordinary mind."
Becky Bace, NSA

Forward by the Author

Sometimes we have to look back in order to look forward. These articles, prose poems, "secular sermons," call them what you will, were written as the digital era gathered momentum. People were flummoxed by the changes confronting them. Most focused on the specifics of their business or organization, but some wanted to know about the deeper implications of the digital revolution.

I wrote the first "Islands in the Clickstream" 25 years ago in response to a request from the Wisconsin Professional Engineers for a column for their magazine that addressed the question, "What is cyberspace?" They liked it and asked for another, then another. Meanwhile the Internet expanded and others requested my column attached to email, a new idea back then. Within a few years they were going to thousands of subscribers in sixty countries. Then Syngress, now part of Elsevier, asked to bring out a collection in 2004--my first book (I just published my sixth, a novel, *Mobius: A Memoir*).

You can buy the original complete *Islands in the Clickstream* via used book channels or from Elsevier as an ebook.

I am omitting most original dates of publication. References to existing technologies and events will tell you all you need to know about the context they addressed. The insights were genuinely prophetic--they illuminated how the digital world was restructuring our thinking, our identities, our opportunities, our entire global society.

The insights still apply. *Islands in the Clickstream* is more than current, it continues to be prophetic.

Take a look. See for yourself.

Richard Thieme
March 23 2021

Table of Contents

Forward by the Author v

Learning to Live in Cyberspace (1996: the first column) ... 1

Ferg's Law 3

Dreams Engineers Have 5

Games Engineers Play 7

Fractals, Hammers, and Other Tools 9

Darling ... Are You Real? 11

Sneaking Up On Ourselves 13

The Air We Breathe 15

Voyagers 17

Densities 19

Beanie Babies and the Source of All Things 21

Whistleblowers and Team Players 23

The Crazy Lady on the Treadmill 25

The Enemy is ... WHO? 27

Life in Space 31

Hacking Chinatown 33

Mutuality, Feedback, and Accountability 37

The Day the Computer Prayed 41

An Owl in Winter: Millennium's End II 43

Night Light	45
Invitation to a Seance	47
A Digital Fable	49
Child's Play	51
Beyond the Edge	53
Detours	55
A Model for Managing Multiple Selves	57
The Next Bend of the River	59
Mapmaker, Mapmaker, Make Me a Map	61
Signatures of All Things	63
When Should You Tell the Kids?	65
In the Crazy Place	69
Contact	73
Why We Are All Getting a Little Crazy	75
Between Transitions	77
Spacetime, Seen as a Digital Image, Already Fading	81
Autumn Spring	83
A Miracle by Any Other Name	85
Talking to Ourselves	87
Coming of Age	89

Learning to Live in Cyberspace (1996: the first column)

A clickstream is the record of every transaction we make on the World Wide Web, that visible and notorious dimension of the Internet we access through a friendly graphical interface called a web browser. Every click of our mouse leaves a trace in the melting snows of cyberspace. Those tracks make up a clickstream, a trail through the virtual worlds we are all learning to inhabit.

"Islands in the Clickstream" will examine ways that life in the virtual world is changing life in general.

The Internet is both a symbol of and a vehicle for the transformation of work and life. A few years from now, the Internet may well evolve into something else—the forms of electronic connection are not inviolable—but the fact of connection is.

The human species is growing a new nervous system.

The Internet does not demand merely a new set of skills; the Net is a new culture, with mores, customs, and its own rules of netiquette. Learning to do business in cyberspace requires patience, attentiveness to how things work in the virtual world, and a willingness to call into question some of our fundamental assumptions about how we do business.

"Islands in the Clickstream" is a travel guide to that new culture. Privacy and surveillance, personal and corporate security, marketing and sales, PR and disinformation, how to avoid net scams—we'll look at all this in future columns. This month we'll look at how we experience change due to the revolution in information technology and what we can do in response.

It is said that someone once asked Ernest Hemingway, How does a person go bankrupt?

Two ways, he replied. Gradually and suddenly.

That is also how individuals and organizations experience transformation. Inside the old paradigm, it feels as if we're learning little by little, adding each new fact onto the last ... until the paradigm snaps. Then we struggle together through a threat of impending chaos to learn new behaviors or new ways to frame reality.

Paradigm change is easy to manage in the seminar room. In real life—where it translates into downsizing, bankruptcy, and new rules for utilities, government, manufacturing—it isn't as easy.

Individuals and organizations everywhere are going through this passage. Radical change is traumatic. It always causes fear, rigidity, and isolation. We've lived with fear, isolation, and rigidity so long. we think it's the normal condition of life. The high level of stress turns our offices and cubicles into zones of "fight or flight." We turn colleagues and customers-the very people who should be our allies—into enemies. We collude to create an artificial scarcity of affirmation and power in the work place. Internal competition undermines our best intentions and drains our energy.

There is an antidote, however: the creation of structures of mutuality, feedback, and accountability.

Every quality program—from quality circles to re-engineering—builds structures of mutuality, feedback, and accountability. This is not an accident. All three are necessary for organizations to remain viable through times of accelerated change. The absence of any one of them skews organizational life in predictable ways, with predictable consequences.

Structures of mutuality, feedback, and accountability keep us flexible and effective under rapidly changing conditions. On the human level, they provide the security we need to keep building the bridges—even in the middle of the air—that we're using to cross the abyss. As a worker in a bank told me, they enable us to build the airplane while it is flying.

New paradigms call us forward through successive zones of transition. The center is constantly shifting. People who had grown comfortable living at the center find themselves on the edge. They need to learn new skills or, at the least, make alliances with people whose skills are relational and whose gifts include creativity and knowing how to make connections. Traditional "outsiders" can help center-dwellers learn how to live on the edge that is always arriving. They can help "traditionals" move out of the rigidity of hierarchical thinking into the more transitory and flexible structures mandated by change.

Paralleling the change in how we work, the revolution in information technology is changing our basic experience of ourselves. Just as the printing press was a catalyst for the Renaissance and the Reformation, life in cyberspace is changing what it means to be a human being.

Networks are subversive of hierarchical structures.

When we connect to a network, we experience ourselves as a point of presence at the center of the web. Our ongoing experience in a web changes how we think about possibilities, how we frame options for action, how we behave.

Persons in hierarchical structures, for example, talk about those structures as if they are external to themselves. They complain about the organization as if it is doing something to them. When they grasp that they are the organization, they move from the edge to the center again.

In a network, there is always more than enough power and affirmation to go around. No-one is ever displaced from the center.

Think of how characters in modern movies change shape, like Jim Carey in *The Mask* or the bad robot in *Terminator 2*. That special effect is called "morphing." It is said that the only competitive advantage a business has today is the capacity to morph like that into new forms.

The same is true of individuals. Only the self-conscious creation and nurturing of structures of mutuality, feedback, and accountability will give us the security we need to risk changing so we can move together in the right direction.

Ferg's Law

This is how the Internet works:

Somebody in Kentucky finds one of my columns and asks to reprint it in a newsletter. Our email exchange begins a dialogue—in this case, on Buddhism, online spirituality, and how the world works—and in one of her exchanges, my email pal says, "I have a friend, Jim Ferguson, and this is Ferg's Law:

When everything can go right, it will, and at the best possible moment."

If that's how computer networks work, that's how life works too. Too many bugs or breakdowns or glitches can obscure the bigger picture, that a vast complex network pretty much works pretty much most of the time.

We are pulled in different directions by conflicting evidence. Jokes abound about differences between optimists and pessimists. But this is deeper than that. This is about the tentative conclusions on which we base the way we live our lives. If the evidence were simple, our decisions would be simple too.

This weekend I am in Iowa for a family funeral. It was funeral weather, overcast and cold. We walked through snow to a hole in the ground and stood shivering in the zero wind as final good byes were said. I have attended dozens, no, hundreds of funerals, and always there is a desire for some sign from beyond the grave. And always there is silence, the answer to our questions is silence, and finally the silence becomes the question to which we must discover our own answers.

Do you ever get sick and tired of the negativity, the whining and complaining about the Internet, computer technology, and where the world is headed? Negativity is a mode of control, a way to try to make people, places, and things fit into a manageable box that we can sit on or manipulate.

I guess we need to live within safe boundaries.

I was diving once in an isolated bay on Maui, far beyond Kapalua, where tourists seldom venture. I was swimming out over the dark corrugated texture of the reef. Toward the mouth of the bay and the open ocean, curtains of blue and deeper blue shimmered in the distance. Suddenly the reef ended and the drop below me was hundreds of feet to a sand bottom. I felt the loss of safety represented by the reef but kept swimming. Then, beyond the curtains of deep blue something moved, something large and dark, so large I didn't know what it was, and the next moment it was gone.

I turned and swam back toward the reef. Once I was over the coral again, fear of that unknown dark form disappeared. The reef represented the safe harbor we are always seeking, while the open water with its unknown possibilities was the invitation of life itself.

It is time to leave our comfortable rooms, the poet Rilke wrote, every corner of which we know, and venture forth into eternity.

Web sites work best that lead us by easy stages from accessible text or images into the complexity of information patterned beyond our comprehension. Our thinking too leads by degrees of precision from simple manageable truths to the highest level of insight.

When the Buddha became concerned that his teaching was at a depth most people would miss, he began salting his stories with "sandbox stuff," the elementary truths we need to remember: Don't hit. Be gentle and respectful. Don't take other people's stuff.

Murphy's law is a true description of life at the lowest level of insight. Things that can go wrong usually do; the tendency to break down seems to be woven into the fabric of all of our projects and woven as well in stars that explode and galaxies that disintegrate. The myth of heat death articulated by physicists, our current high priests of cosmology, turns the silence of the grave into the silence of the universe.

Standing at the grave, I remember a friend, an artist named Jim. He called me from the hospital to say he was dying, but before I could visit, he checked out for a final European trip with his mother and sister. In a tourist hotel in London he lay down and died. They shipped his body back home for burial.

The following week I was discussing plans for a memorial service with his companion. Now, Jim always had long wild hair, shoulder length hair, in keeping with his artist image. As I spoke with his companion, there suddenly emerged on the edge of my consciousness like a stained glass window brightening as the sun came out from behind a cloud Jim's face, and it stayed there, unlike a memory, as the conversation continued. But for whatever reason, his hair was very short.

Then he faded, and moments later, his friend mentioned that—to please his mother's conventional sensibility—he had cut off all his hair before leaving on that trip.

We can explain that event at any level of precision, but whatever our interpretation, something emerged in my consciousness that told a more precise truth than we usually know how to tell.

Negativity is a way to build a dark familiar reef under our swimming selves. The ultimate source of negativity is a lack of courage and a need to make the darkness safe, rather than risk the open water.

At the graveside—and we are always at the graveside—the powerful compression of grief tunes our awareness to what matters most. We surrender to the truth that is always there, but buried, our deep longing for forgiveness and mutual forbearance, our desire to surrender the need to be rigid or right. The readiness is everything, and during those moments of exquisite timing—tolled by a clock that ticks to a different rhythm—we know that when everything can go right, it will, at the best possible moment. We weep, and we embrace one another. The universe is gregarious and welcoming. We are built to live in space that is gateless, unbounded, free.

Dreams Engineers Have

I confess: I'm a right-brain guy in a left-brain world. Images and visions are more real to me than abstractions, I see the future more easily than things that are right in front of my face.

That's why I started writing fiction and sold my first story at seventeen. "Pleasant Journey" was published in *Analog Science Fiction*. It concerned a man selling a virtual reality machine to carnivals. Attach the electrodes and off you went into your own dream world. The carnival owner tried it out and didn't want to come home. He wanted to stay in that virtual world forever.

I studied liberal arts. We were taught that art and literature mattered most; the loss of an art object or literary work was a tragedy. I remember a professor weeping for the lost plays of Aeschylus.

No one grieved, however, for the streetlights of Cordoba or the sewers of ancient Rome. Engineers were practical people. Their plans and drawings were seldom the subject of scholarship, and I don't recall a single course in the art of engineers and how their dreams made real the infrastructure of our civilization.

In part, that was because plans and drawings were never intended to last. Once pencils were invented, plans were sketched in a way that smacked of impermanence, like something you'd draw on a napkin over lunch.

Leonardo da Vinci filled his notebooks with plans and sketches. Those notebooks, detailing his dreams, nearly disappeared after his death. He never published their contents, and more than thirty volumes were left to his friend Francisco Melzi with instructions for printing. Instead, they were ignored for fifty years. When the contents were finally published in 1880, most of Leonardo's inventions were obsolete.

Bill Gates paid a small fortune for those notebooks. He knows that they're works of art worth owning—the dreams that prefigure our civilization.

It was no accident that my first short story was science fiction and concerned technology enabling us to transform our lives. That's the story of our century. The invention of electronic media, including the Internet, is the infrastructure that enables dreamers and thinkers to be creative in new ways. The medium is so much the message that we're writing stories about the technology rather than the life it enables us to live. That will change, though. The technology, the new media through which we express ourselves, will fade into the background and become as transparent as contact lenses.

Henry Petroski's magnificent study of *The Pencil* begins with an anecdote about Henry David Thoreau. Thoreau made a list of everything he needed to take to his life in the woods but neglected to mention his pencil. Yet his pencil was always in his pocket, and the Thoreau family business was ... making pencils.

Science fiction is the way men and women in the twentieth century have dreamed of the future. I like to joke that people who call me a "futurist" are mistaken, that I describe the present to the ninety-five per cent of the population that hasn't arrived at it yet. That's why it sounds like the future. It's the same with science fiction, which depicts what is right in front of our faces, coming around the corner at the speed of light. It sounds like the future only if you aren't noticing what's happening.

I recently did an article on biometric identifiers—retina and iris scans, fingerscans, voice prints, and the like—the use of part of us to stand for all of us. Those digital artifacts don't merely stand for us, however, they become us in the social, economic, and political worlds to which they allow or deny access.

Our word will not be believed when a retina scan refuses to allow us into a secure area, just as we used to say "photographs don't lie," believing the photo rather than the person in the photograph. Now that photos and all forms of information can be digitized, we know that photos do lie. A photo is no longer worth a thousand words when both words and images are subject to digital manipulation.

What will it be like in the virtual world in which digital bits "pass" for ourselves? Let's go further. What will it be like when the information that IS ourselves—i.e. our DNA code, the drawing or blueprint that is expressed as our bodies, our minds, our lives—can be uploaded and stored?

Teleportation used to be a sci-fi subject. Two years ago, an international group of six scientists confirmed that perfect teleportation is possible—but only if the original is destroyed.

That theoretical work changes teleportation from a sci-fi scenario into an engineering problem. If the information that constitutes our pattern or code can be transmitted and replicated, and the original is destroyed in the process—who arrives? Who is left behind?

In a similar way, we used to think the hard copy was the "real" document and photocopies were secondary. Now we think the virtual copy stored in digital memory is the "real" document and hard copies are mere images of the real one.

The network is the computer, and Marvin Minksy reminds us in "The Society of Mind" that turning over a multiplicity of representations in our collective mind instead of getting stuck in one way of seeing things is what we mean by thinking. The network does the thinking. We are merely cells in a single body, and a human being alone—like a stand-alone computer—is a brain in a bottle.

A Zen monk held up a cup and asked what was most important about it. One pupil said the handle, another the bowl, but the monk shook his head. "The most important thing about the cup," he said, "is the space it creates."

The Internet is "space" brimful of possibility and potential, but by virtue of its structure it organizes the form of our thinking and dreaming. Engineers who build the infrastructure of the world create the space in which we live and move and have our being, and we don't even notice. It's as transparent as Thoreau's pencil. We don't even know who's dreaming any more—the individual or the collective mind—and what is science fiction or science fact. We DO know that engineers dream up our space and, like God in creation, are everywhere present in our lives but nowhere visible.

That's the cost of making sketches with pencils. That's the cost of using materials that decay. But then, everything decays, and digital images are more transitory than drawings. Art and artifact converge, and those who build the infrastructure that informs how we dream are at least as creative as Aeschylus, as practical as Leonardo, and as holy as that Zen monk.

Games Engineers Play

Our societies teach us the skills we need through games. Playing games is how we explore possibilities, identify talents and interests, learn values.

Computer games are exploratory toys for investigating the digital world, but more than that, computers themselves are toys. The games—how they are built, how they change us when we play them—are a catalyst in the evolution of new ways of framing reality.

This summer at DefCon IV, the hackers' convention, I met a hacker who was twenty years old, an "old man" of hacking who had retired to mentor younger hackers.

He had been programming computers since he was six, when he programmed Pong on a computer that had no long-term memory. He remembers the sense of loss when, after weeks of Pong-playing, he turned it off and lost his beloved binary companion.

That young man is in the vanguard of a new variety of human being. I call them *Homo sapiens hackii*.

Culture is a human simulation of genetic evolution. Our cultural symbols transmit what we learn to new generations. But we are changed by the structures of our cultural forms as well. The structure of our information systems determines how we think, how we frame reality. Our symbolic structures engineer our psyches in their own image and determine how we hold ourselves as possibilities for action in the world.

That hacker and his generation are trained from infancy to understand the world in ways dictated by computers. From one point of view, human beings are transitory forms for organizing and disseminating information, and the form of that organization is determined by our symbiotic relationship with our symbol-manipulating technologies, in this case, computers.

Thirteen years ago I bought an Apple II+ computer. I began playing with LOGO, a child's version of LISP. I learned about recursion working with LOGO and I learned how to write a program to substitute verbs and pronouns for the ones typed by a user.

Joseph Weizenbaum at MIT had just used LISP to wrote ELIZA. ELIZA is a simple substitution program—a natural language parser—that enabled a computer to simulate the responses of a Rogerian therapist. Such therapists repeat back to the client what they think they have heard, thus returning responsibility for their own thoughts and feelings to the client.

People playing with ELIZA, however, responded as if they were engaged in an intimate conversation with a real person. Weizenbaum was upset and wrote one of the first "dire warning" books about the future of computers. The power of the computer to elicit projections was so strong that he felt we were in danger of losing our souls to the new machine.

Games like ELIZA taught us to project a gestalt or complete personality onto a machine that mimicked the response of a human being. Playing with ELIZA taught us how to relate to GUIs, expert systems, and smart agents. We were learning to relate to our projections as if they were external to us in order to interact easily and effectively with complex computer applications. We were learning a technique for manipulating symbols that tamed the power of the machine and made it manageable. This is analogous to playing word games as we learn to read and think in text.

A natural language parser, exchanging words and seeming to respond intelligently, generated a new genre of interactive fiction. The classic text games of Infocom stand out as the best.

One day my son and I were playing *Hitchhiker's Guide to the Galaxy*. I had written fiction and taught English literature, so I understood how text worked. Symbolic textual narratives disclose horizons of possibility far into the distance, toward the horizon of the text.

Playing that Infocom game disclosed a different kind of horizon, a different set of possibilities. The structure of the game itself, determined by computer programs using recursion, changed me as I played it. The game created a new way of framing myself. I learned to imagine my psyche—and my life—more as a recursive fractal landscape than a straight line.

Of course, a single game did not do all that at once. But playing computer games and experiences like that did disclose new possibilities for being human. The structure of that game is identical to the structure of the Internet. Networks, like fractals, are self-similar at all levels. Networks of networks look like networks. Interacting with networks changes how we think.

Newer games using movies and 3D-VR do the same for visual space. One of my recent favorites is *The Pandora Directive*, an interactive mystery. One of the "game spaces" is a virtual board room with beautiful indoor trees. As I watched the screen, my hand rolling the mouse across the pad, I noted the smooth glide of the trees past a large window. I turned and looked back at the green leaves and the gleaming wood of the conference table. Then I rolled out into the hall.

The following week I walked into a boardroom in a bank building that had similar trees, a similar table, a similar window. I felt myself rolling through the real room. The perceptual framework through which I experienced the real room was an image of the virtual world.

These days, reality often imitates a simulation, rather than vice versa. Playing games like *The Pandora Directive* creates the psychic space through which we subsequently filter our experience. The computer program programs us.

So look to electronic games for signs of what's coming next. Children play *Doom* while their parents guide smart bombs to bunkers in Iraq. Adolescents build robots while their parents use remote vehicles and telepresence to explore radioactive "hot spots."

On ESPNet SportZone, there's a virtual world of sports. Web surfers use data from the site to create fantasy teams. They channel aggression into fantasy football or baseball, a harmless pastime that helps bleed excess energy from a civilization with too much time on its hands. They do this by substituting images and symbols for the real thing.

But then, what else is civilization, but the weaving of a web of images and symbols which we mistake for reality? The first spoken and written words did the same. The game has not changed, but computers take the game to a higher level.

Fractals, Hammers, and Other Tools

"Fractint" was one of the first computer programs I encountered that blew my mind. (It's still out there on the Internet. Download one if you want to try it.)

Fractint generates fractals. Fractals are mathematical formulae that express complex realities with elegant simplicity. Before computers, you had to have a mathematician's mind to grasp the relationships expressed by fractals. Computers enabled those relationships to be represented pictorially. Fractint lets you generate images of fractals, then cycle through them in thousands of colors. The vision of a fractal in action is stunning.

Fractals often resemble natural objects. Simple formulae using recursion generate images that look like branching trees, clouds, coastlines, or fern leaves. Seeing those images on a computer changed how I saw the natural world. The computer generated a different framework for looking at and comprehending the "real" world.

Fractals are self-similar at all scales. If you magnify a section of a fractal, then magnify a section of the section, each one looks similar, like nested wooden dolls. You can keep magnifying smaller and smaller pieces until the image on your monitor is part of something so big that, if you spread it out, it would stretch from the sun to the orbit of Jupiter.

My wife, who is not a geek, looks up now as we walk through a forest or watch clouds move through the sky and says, "Fractals."

This brave new tool, the computer, is programming us to see things in its own image, teaching our minds as well as our mouse-clicking hands how to use it.

Fractint also taught me that intellectual property, as we have known it, is over.

The concept of an "author" who owned "a work" was invented by the printing press. The printing press fixed words in text and created an illusion of permanence, of something solid "out there." Students are still surprised to learn that Shakespeare did not care to preserve his plays for future generations. "Writing for future generations" was a conceit thinkable only after we had fully internalized the world of text and thought of books as artifacts that would last.

Fractint was built by "the Stone Soup Group," programmers who worked collaboratively online. Some of their names are known, but many are anonymous. A collective wrote the program, just as monasteries in the middle ages created illuminated manuscripts without a thought for the name of an "author" or owner of the "intellectual property."

Cultural artifacts like laws (copyrights, patents) are tools too. The shape of those tools is determined by our information systems. After we use them a while, we forget that, and they become part of the background noise of our lives.

Fractals are a metaphor not only for what I see "out there" but also for what I observe within myself. Every decade or so, I discover myself in transition to another developmental stage. Each stage includes and transcends everything that came before. My psyche is self-similar at all scales, just like a fractal.

Civilizations too go through developmental transitions, and they too include and transcend everything that came before.

Back to tools.

It is said in the consulting business that "to the person with only a hammer, every problem looks like a nail." Our tools structure our perception and frame our possibilities for action.

I asked a number of engineers what tools they commonly use. All but one said "computer" first. Some added T-square, or architect's rule, or drafting board. Only one said pencils, although everybody uses them. Nobody said "words."

We only notice the new tools in our kit, like computers. Those we were given by prior generations disappear into the background. I notice that most people mean by the word "technology" the technology that has been invented since they were children.

The evolution of tools and the hands that hold them or the minds that think them is a cultural process. It's a chicken-and-egg kind of thing. Did we build more complex bridges and buildings because we had better tools, or did tools evolve that enabled us to build better bridges and buildings?

Computers simulate what we call "reality" but that "reality" in fact consists of nested levels of symbols. If it were a mathematical formula, it would look like this: Digital images => printed texts => writing => spoken words. They are all artifacts, nested in levels of abstraction that are self-similar at all scales.

Before human beings spoke, the artifacts or tools generated by language did not exist. We call those tools ideas, concepts, mental models. They are the building blocks of our maps of reality. Because they are modular, we can connect words and ideas in an infinite number of ways and build more ideas, more elaborate frameworks or architectures that enable us to build everything from bridges to religions.

Like speech, writing, and print, the computer is a tool that shapes our perceptions into forms the computer can use. If we are to bring our ideas to the computer, they must be expressed in language the computer understands.

To the person with a hammer, everything looks like a nail. To human beings who use speech, the only ideas we can think are ideas we can express in words. In a civilization transformed by interaction through networked computers, we will think only thoughts that can be simulated or manipulated by the worldwide network that mediates communication and the flow of information.

The world looks to me like fractals because Fractint taught me to perceive the world as fractals. Engineers will build the kind of infrastructure that networked computers teach and enable them to see and think. The physical structures of civilization will be determined by how computers think.

Everything is a flowing, the Greek philosopher Heraclitus said. If only he'd had a PC and a program like Fractint! Then he could have seen that flowing in thousands of colors, fractals of unimaginable simplicity and complexity, self-similar at all scales.

I bet it would have blown his mind.

Darling ... Are You Real?

It is not news that sex sells. Nor that new media often contain sexual images. The first books, the first photographs. The initial demand for VCRs in the home, creating a critical mass that enabled Hollywood to sell films. And, of course, the Internet and other digital media.

Because pornographers routinely shoot scenes from various angles to meet the needs of diverse markets, their works are a natural for Digital Video Discs. Users can click on a scene from as many as sixteen angles and zoom in and out. It's an interesting mix on the user's part of a need for control and a need for the person or image they manipulate to be or seem to be real.

A "cyberbabe" sits in a cubicle responding to typewritten commands from people paying four dollars/minute to interact with her image in a small square on their monitors. Writing for Wired, Frank Rose wondered what request the women received most. It was not an explicit sexual act but ... "Would you please ... wave?"

Users need to know that the tiny image of the dancing digital doll ... is real. They want her really to be there ... and to be there just for them.

The mind boggles. The human soul is so hungry for self-deception that it will swallow a pig whole.

Those who manufacture pornography are not known for originality. Quite the contrary. Their clients do not want innovation or creativity, they want predictability, and as a genre, pornography is nothing if not predictable. Scenes follow a formula—A x B x C x D x E—the letters corresponding to increasingly explicit scenes, the Xs corresponding to filler that lets the user inhale before the next escalation. That's why, for example, Vladimir Nabokov could argue that Lolita was not pornographic.

The requests that flow to those booths are easily anticipated. Like children who never tire of hearing the same story read the same way. and who in fact will object if a single detail is different, clients want the reassurance of the same scenario played out the same way again and again. The comforting touch of a mother as much as a lover.

It ought to be simple for programmers then to set up a database of video clips indexed to specific requests, the variations linked to a natural language interface. When the client types, "wave your hand," the clip of a lady waving will respond. A hundred and fifty waves with each hand ought to handle the most suspicious client.

The next step, of course, is not to have an actor in the box at all. Just as session musicians who counted on studio work for their livelihood have lost jobs to synthesizers, a digital cyberbabe can be constructed, pixels of light substituting for an image of a real person. It isn't the pixels but the pattern, after all, that makes an image a symbol, and it's a symbol that clients of a digital interface—any digital interface—need.

Come to think of it, we used to call them "movies" ... images of shadow and light simulating the appearance of real people so well that we had to keep saying to ourselves, "it's only a movie" when we grew afraid or developed a crush on a leading man or woman—someone who wasn't there, never had been there, someone who hired a staff to pretend to be them when we wrote to confess our love.

People used to think characters in novels were real too. The Dickens character, Little Nell, was followed by serial readers the way soap operas are watched on television. A crowd burst into tears when the captain of the ship bringing the latest installment to New York told them that Little Nell was dead.

The film noir masterpiece *Body Heat* depicts a lawyer deceived by the words of the women he loves. He says at the end—speaking perhaps for all of us—"experience shows that I can be convinced of anything." We have to pinch ourselves to remember that there's no there there, that Mattie Walker is only a digital image of someone pretending to use speech to deceive.

The best "hacking" is done by people working on people, not computers. It isn't data that's relevant but patterns constellated by data, the intentions of a person in a particular context revealed by the way the data seems to connect. "Social engineering" is the art and craft of eliciting information by pretending to be someone you're not, acting a role that blends with its surroundings so seamlessly it seems real.

Which is one reason why businesses have replaced countries in the post-Cold War global free market, why intelligence and counterintelligence, information and disinformation, are axiomatic to remaining viable in a knowledge economy. The local head of the FBI just joined an accounting firm to work on fraud. A well-known hacker joined a big six firm as head of their Tiger Team. Business, intelligence, and hacking are in many ways indistinguishable.

It must have started with speech. One can imagine the shock when speech emerged in human culture and people realized that someone could know their thoughts merely by ... speaking with them, asking questions. Speech must have become a means of hiding in the same instant it became a means of self-disclosure. Truth and lies are Siamese twins, joined at the lips.

So the difficulty of knowing—"darling, are you real?"—is not new to the digital era. Nor the celluloid era of Meg Ryan as Sally in that restaurant scene. Our experience does show that we are capable of believing anything, that our primitive brains take appearance at face value. Betrayal and self-deception as a cause of human tragedy are as old as story and song.

And so ... I want you to know that I am here, I am real, I am saying good-bye ... and waving my hand toward the monitor as I fade into the distance ...

Sneaking Up On Ourselves

It's pretty tricky, sneaking around a corner which is really the surface of a sphere until we are looking at the backs of our eyeballs with our own eyes. That's what happens, though, when we see ourselves seeing ourselves.

That may sound philosophical or even religious, but we really have to think like this at the dawn of genetic engineering. The practical applications are immense. We are beginning to engineer our very selves, to see our subjective ways of framing reality and being in the world as modular objects to be manufactured and enhanced. In the next century, it's going to be bootstrapping big time for the human species as we see ourselves as an image in our own collective mind, an image to be designed and executed. Not exactly a computer program, but something very like.

That's a metaphor, of course, and metaphors are horses we can ride only to the limits of out thinking. Then they dissolve into thin air, and we ride on horseless like cartoon roadrunners racing off cliffs into the middle of the air.

Subjectivity is hardwired into our brains. That's a way of saying that reality is an illusion, that even thinking we can slip outside the limitations of our minds—an experience that Zen adepts call enlightenment—and see what we are and what we are not, that is, see ourselves from a point of view outside ourselves—maybe that's an illusion, too.

But think about this. A patient with a brain disorder was being probed with electric needles. Doctors were hoping to locate areas of the brain that were the source of her seizures. When the tip of the needle touched a particular area, the patient began laughing, laughing not at the needle or in response to it, but laughing enthusiastically and spontaneously because … well, because they were all so funny.

Funny? asked the doctor. What's funny?

Well, she laughed, looking at them with sparkling eyes … the way you're all just … standing around … is … so funny!

I remembered a friend experimenting with hallucinogens. The drug induced a chemical change that enabled him to see clearly that everything in the world was … show business! He couldn't stop laughing at the spectacle of role-playing human beings acting as if their social masks were real. Whatever anybody said, he cried out, "That's show business!" and dissolved into laughter once again.

Another time he saw the earth as a marble, like the photograph of our planet from the moon or the world seen as a walnut in the hand of her Lord by Julian of Norwich in her anchorite's cell. He saw the earth as a sphere from the center of which a million lines radiated to the surface, at the end of which walked tiny ant-like humans crying, "Mine! It's mine!" Again, he laughed hysterically. And who can blame him? Humans, part of creation, acting as if they owned abstractions like "land" … acting as if their consensual hallucinations were reality instead of a game they agreed to play.

Stephen Hawking gave a lecture last week at the White House. Through the "wonders of modern technology," a RealAudio/Video feed played on my desktop as I worked late into the night. His electronic voice said that, unlike Star Trek or Star Wars, in which contemporary humans are projected pretty much as we are into the distant future, he believed we would recreate ourselves through genetic engineering in ways we can not even imagine. People think of obvious things—greater strength, longer lives, faster reactions—extensions of how we are today that make us more and better, but not significantly different. The real differences will be created and discovered as we learn how laughter and sadness happen, how we construct reality, how we use mood-altering experiences like religion, therapy, entertainment, other kinds of cultural play, how we mean and be.

Some religious sentiments—like wonder and awe—have been linked to particular genes, just as parts of the brain, altered by the use of spiritual tools, ingestion of chemicals, or electric probes, spontaneously generate visions that are ... so funny! Visions of human beings acting as if we are gods on a planet from the oceans and mud of which we have only recently crawled. This civilization is a phase, not a conclusion, and this earth is just one of the millions of planets teeming with life that will teach us how we have evolved and maybe even why.

Zen masters disdain the use of chemicals or electricity to induce enlightenment. They prefer traditions handed down for generations. That perspective may soon seem quaint and archaic. Genetic engineering will enable us to say at a particular level of description—a level appropriate to biochemistry and biomechanics—what happens when we experience that which emerges in our subjectivity as awesome, wonderful, or funny. We will create emotional states of being that do not yet have names. The beings that feel, think, experience those states—our intentional progeny—will embody that unimaginable next step for the species of which Hawking was speaking.

Ethical reflection never anticipates the breakthroughs that make us sit up straight, nor do legal niceties. What will we do when our mistakes exhibit behaviors we don't want? Put them in reservations or safari parks, use insect-like cameras to scuttle among them, amusing ourselves with their wacky antics?

Will they vote? Will anyone? And will it matter?

Self-transcendence is an emergent behavior, perhaps the beginning of a rising spiral of possibilities: The ability to see ourselves from a meta-perspective, sneaking up behind ourselves and watching ourselves think thoughts like these. The ability to experience the disappearance of what we call our "minds" as we realize that "no-mind" simply describes what's so. The ability to decide what's funny or whether to include humor at all in our genetic program ... some of our progeny roaring with laughter at the mere thought while others stare uncomprehendingly, wondering what's so blanking funny.

The Air We Breathe

Nothing is harder to see than what we believe so deeply we don't know we believe it. That's why a frontal assault on our core beliefs is always doomed. Our minds think they themselves are under assault, rather than the beliefs they have adopted, and defenses go into gear to rationalize, minimize, or deny what they're hearing. Or else the anomalous data creates so much cognitive dissonance that our minds just plain shut down.

The degree to which technologies of communication, surveillance and control have insinuated themselves into our everyday lives is striking. Here in Wisconsin, a bill just sailed through the legislature that expanded the state's authority to collect health care information. The bill allows the Office of Health Care Information to collect and publish financial and other data from doctors and health care providers in addition to data gathered from hospitals and ambulatory surgery centers.

Remarkable to those concerned about "function creep" was the lack of concern on the part of the public. Everyone pretty much lined up on behalf of "efficiency and safety," the two horsemen of the apocalypse of privacy rights. The legislative committee was "stacked" on behalf of the measure and the public was informed after the fact, the bill having been called suddenly the night before the vote was scheduled.

This is a holographic slice of a bigger picture. The technologies of linkage and the power of those who profit from using them are the true weapons of information warfare. That war is fought not with lasers and satellites patrolling the "high ground" of earth orbit, but in the trenches of our daily lives. Because the consequences of ubiquitous linkage are often invisible, the average person—with limited time and mental resources—is unaware that the hidden infrastructure of a global political economy is being built out of the mundane data of their lives.

When I recently pressed a career officer in the intelligence community about practices that alarmed me, he maintained that those practices were illegal, hence nonexistent. After a few drinks, however, he acknowledged that many intelligence agents find it easier to ask forgiveness than permission and act accordingly. That all-too-human reality is why we will pay in the future for every time we refuse to speak or act in the present on behalf of the privacy that secures our freedoms. Without secure boundaries, there are no individuals ... and no individual rights. The primacy of the collective, a by-product of the transforming power of information technology, is paradoxically entering mainstream thinking as a priority through the political action of those who believe they are supporting a conservative, business-friendly agenda.

It's as if the entire world is joining NATO, justifying Cold War behaviors by invoking the Evil Enemy. But unlike the Cold War, when there was at least another camp, the "other side" now means people anywhere who oppose the converging self-interested policies of the military-industrial-information complex.

And now for something completely different.

Children's toys are often an early warning system in which the future first becomes visible.

"Sound Bites" is the name of a new technology recently introduced at the annual Toy

Fair in New York. A person inserts a lollipop in a Sound Bites holder, and when they bite into it, sound vibrations travel through their teeth to the inner ear where they are heard as normal sounds. This magical effect lets snackers hear music (guitars, drums, or sax), special effects, or voices.

The notion of slipping advertisements, propaganda, or suggestions into our meals is so outrageous I expect it to be adopted without a murmur. One imagines voices coming into our heads from every artifact. Deserts in the company cafeteria, basketballs as we dribble down the court, even sex toys will all have something to say. Everything will be a means for communication ... as indeed, everything already is, but today those messages are still mostly implicit, while these songs and jingles will be as explicit and close to our noses as bumper stickers.

And now for something even more different.

A hobby in which I have indulged myself for years is the investigation of UFO phenomena. It's an interesting puzzle, requiring cross-referencing texts in the public record with the confidences of mostly plain people, as well as intelligence agents, air force officers, and airline pilots. Like most amateur investigators, I find that ninety per cent plus of what I read or hear can be explained or discarded, but—again, like most—the remaining accounts are pretty compelling.

Yet what interests me as much as the data is the widespread ridicule that greets even the most reasonable statements about the phenomena, e.g. it is worthy of investigation, if only as a psychological or sociological phenomenon. One hesitates even to mention this interest because of that predictable response.

Such ridicule apparently became official policy around 1953. Before that, for five years (1947-1952), UFO phenomena was taken seriously by governments in public and private. An early head of Project Blue Book stated that behind the Pentagon's closed doors, the argument was not about the reality of the phenomenon, but whether its origins were Russian or extraterrestrial. A widespread wave of sightings in 1952 became the point of departure for a policy of debunking. Air force fighter pilots and commercial airline pilots alike have told me how they and their colleagues learned quickly not to risk their careers or reputations by making a report or going public with details of an encounter.

Indifference to the erosion of privacy rights ... candy that sings to our brains ... a policy of public ridicule that discredits innocent people.

It is easier than ever to engage in sleight-of-hand, manufacture a consensus, and manipulate dissent. Yet the truth too is boosted by technology. Truth too sings to our brains, and the linkage technologies that magnify the fictions we seem to need to sleep easily in our beds will disseminate as well the truths that fuel our hunger for knowledge and our passion to be free.

Voyagers

When we come to a new place or enter a new environment, the landscape looks all of a piece, and we have to learn how to see it in depth and detail. Our interaction with new cultures teach us over time how to understand them.

When I moved to Maui in the eighties, I lived about thirty feet from the ocean. From the sea wall, when I looked out at the channel that defined the "pond" among our islands, all I could see was ... water. Nothing but water. Of course I could see the other islands and waves breaking over the reef, but the ocean itself looked like nothing but undifferentiated water.

Last week I returned to Hawaii to speak about spirituality and technology. During the week, two canoe builders from the Marshall Islands made a public presentation and shared some of their lore, long believed to be lost. "Some of our navigators are trained to detect six different swells by the feel of the canoe," one said. Others specialized in navigating by the stars or weather. Despite the repression of their culture (and the explosion of dozens of nuclear bombs on their islands), they had somehow kept alive the knowledge of their ancestors.

They used tangled knots to map the night skies and learned to discern the subtle interacting patterns of the swells by crouching in canoes on what looked to a "mainland haole" like a perfectly calm sea. When they looked out at the ocean, they saw a lot more than just water.

After a few years on the island, I could see at a glance the direction of the wind and the complex pattern of the currents. That told me what I was likely to encounter when I went diving or spear fishing. I knew where the fish would be feeding, how to co-exist with morays and reef sharks, how to use the surge of the sea to slide without effort toward prey. The angle of the sun under the water, the length of the seaweed, the Kona wind, all correlated with the feeding habits of the fish on the reef. The sea resolved itself into a complex, richly detailed environment.

What I learned was child's play compared to the intimacy with which islanders know the ocean. Someone who lives on the mainland might look at the water and see only a barrier, whereas islanders see an open invitation, a whole world waiting to be explored, both highway and home.

The journey into ourselves and the journey into the symbolic landscape that defines our culture—of which the Internet is an emblem—are the same journey.

When we first turn inward, the landscape may seem opaque, but as we explore through meditation, prayer, and other disciplines, we too discover both a highway and a home.

It's as if someone who spent his or her entire life on land hears for the first time about the ocean. The word calls forth an image and a desire, and that is the beginning of the journey. We make our way to the water's edge and look out at the singular immensity of it. Some plunge in; others take scuba lessons, letting others coach them. When we first snorkel or dive on a tropical reef, we are amazed at the beauty and variety of living forms. That beauty can be a trap. If we're not careful, we stay at that depth instead of learning to go deeper. If we do go deeper, new worlds are disclosed, new possibilities for communion with ourselves, others, and the universe of which we could not have dreamed.

Over time, we become as comfortable under water as on land, and our framework expands to include the sea and what is under the sea as well as the narrower life lived in the air. We move back and forth between them easily. Air and water become dimensions of a single reality.

The first explosive photos taken by the Hubble telescope showed the richness and complexity of space, a technology disclosing new possibilities for action. As those possibilities percolate into our consciousness, what it means to be a human being is transformed.

That's how it felt too when I downloaded my first browser and tumbled like Alice into cyberspace, emerging from underground eight hours later, oblivious to the passage of time.

The Internet is a vast sea of possibilities, a symbolic representation of our collective consciousness and our collective unconscious. When we explore the Net, we are exploring ourselves. The Net is a swirl of invisible currents. We learn to surf swells of meaning that surge back and forth like the sea. We learn to follow currents of information, feeling the swells interact in subtle and complex ways. We become voyagers in the sea of information in which we are immersed, plunging through high seas in outrigger canoes. We make our own tangled star maps that represent and remember for us how to find our way home.

There is ultimately only ourselves to know. When we try to understand everything, we do not understand anything at all, observed Shunryu Suzuki. But when we understand ourselves, we understand everything.

The Internet is not so much a set of skills as it is a culture. Guided by mentors, learning like wolves to hunt together, we learn how to hang in the medium. The images on our monitors are icons, windows disclosing possibilities far beyond our home planet. Inner and outer space alike are explored by tele-robotic sensory extensions, revealing the medium in which we have always been swimming. Consciousness is the sea, and the sea is all around us. The secrets that we think are lost are simply waiting to be found: Supra-rational modes of knowing. Connection and community of such depth and complexity that we grow giddy with delight. A network in which we are both nurtured and fulfilled, each node of the web a reflecting facet like one of Indra's jewels, reflecting each of the others and the totality of the whole.

Densities

Steven Hawking noted in a netcast from the White House that the next generation of humans will live inside a common sense world of quantum physics the way we have lived inside a Newtonian landscape. "Common sense" is simply what we're taught to see, he said, which is why new truths always appear at the edges of our thinking.

Or, as George Bernard Shaw put it, "All great truth begins as blasphemy."

Is it any wonder we are all beset by "cognitive dissonance" and see our reality-frames flickering the way clairvoyants (excuse me, "remote viewers") see images of distant sites? One moment we are living happily inside Newtonian space, walking down a straight sidewalk toward a right-angled corner when ... poof! ... with a puff of smoke, we experience ourselves bent along a trajectory like light pulled by an immense gravitational tug. Then we remember that how light bends IS gravity and what we thought was a "pull" is simply the topography of energy wrinkling and sliding into whorls of various densities.

In a museum the other day I watched a marble spiraling down a funnel of smooth wood, circling toward the vortex. I thought of light traveling along the curves and bumps of space-time ("the universe is shaped like a potato," Einstein said, "finite but unbounded.") I thought of gravitational lenses, created when galaxies that are closer to us magnify and distort more distant galaxies.

Einstein predicted sixty years ago that a massive object would bend and intensify light, generating multiple images or stretching an image into an arc. When everything lines up just right, the distortion becomes a perfect circle, like the galaxy pictured last week in *Science News* (Vol. 153, No. 114).

That's the long view. Turn the telescope around to see what's happening right here in our own digital neighborhood.

Web sites are best characterized not by size but by density. A map of cyberspace would look like millions of galaxies and a map of the traffic between sites would look like a photo of electromagnetic energy across the entire spectrum.

A browser is a knowledge engine that organizes information in flux so it appears momentarily frozen. A site such as Yahoo that links links is a kind of gravitational lens that boosts distant clusters into the foreground. If we could see ourselves interacting in cyberspace, we too would look like energy pouring through our monitors and moving at the speed of light toward densities around which our interests coalesce. Our monitors like worm holes let us bypass the long way around.

Organizational structures, including web sites, are dissipative structures like whirlpools that retain their shape while exchanging energy and information. Humans too are modular structures of energy and information that interface over the Internet. That map of the energies of cyberspace is really a map of our Mind.

Not quite common sense yet, is it? Words slip, slide, decay with imprecision, T. S. Eliot said of his efforts to fix in poetic form the world he discerned. In the world of printed text, the illusion that words and meanings are fixed is magnified. The same words in pixels are obviously transitory. Our media too function like gravitational lenses, magnifying meanings intrinsic to their nature. The digital world builds a "common sense reality" congruent with the quantum world, communicating by its very nature that words, meanings, and all things slip, slide away.

We build this island for ourselves in the always sea and comfort ourselves with the illusion that we are on dry land.

The trajectories of the energies of our lives—how they are organized, aimed, and spent—are determined by our deepest intentionality. How we intend to live our lives is how we wind up living them.

Cyberspace is a training ground for learning to live and move at the speed of our minds, the speed of light, to inhabit a landscape that morphs or changes shape according to our will, intention, and ultimate purpose.

The "sites" in our minds grow denser when our intentions coalesce like millions of marbles rolling simultaneously toward a single vortex. Space, time and causality may be woven into the very fabric of our minds, as Kant said, but in a quantum landscape, causality is a very different animal. An effect can precede its own cause.

Which is exactly how our minds operate.

Consciousness is always consciousness for or toward some end, always an arrow aimed toward a potentiality or possibility. As a mental construct, the image comes first. The effect precedes the cause and causes the effect to come into being. That's why some think consciousness is the origin as well as the goal of evolution.

A recent reflection on maps, filters, and belief systems brought from a reader an account of the moment he realized how much the Mercator projection exaggerated the size of the European community. He recalled the first time he looked at Buckminster Fuller's Dymaxion map that looks at the world from the North Pole rather than the equator. From that point of view, the world is seen as a single unified landmass. The world has never looked the same to him since.

Consciousness manifests itself in a visible medium like the Internet so we can see it. We can never see the thing itself, because there is no thing there. Nothing. But we can see some of the infinite ways it manifests itself. Working and playing on the Internet is one way to practice handling ourselves in a quantum world that is fluid, modular, and interactive, a trans-planetary world, a trans-galactic world emerging on the edge of the grid in which we have been living. That grid contained reality in nice neat boxes. But the grid is flexing, morphing like an animation even as we look at it, turning into another of its many possibilities. Seen, of course—it's only common sense, isn't it?—from just one of its infinitely many points of view.

Beanie Babies and the Source of All Things

Since this column is read in many countries, let me explain "beanie babies" to those who may not grasp their importance.

Beanie babies are toys, cute little animals with cute little names. They're manufactured here in the upper midwest, at least the real ones are. They're good examples of how to create artificial scarcity in a consumer economy. Some beanie babies are hard to find, so Peanut the Elephant or Chilly the Bear will sell for a thousand dollars.

Beanie babies illustrate the dangers of living in a digital world as well as a consumer society. When the value of a commodity or an idea is manufactured and therefore arbitrary, artificial demand inflates prices and invites fakes into the marketplace just as it does in the world of ideas.

Beanie babies are cute enough, but easy to fake if you don't mind inferior fabric or stitching. It's all in the name, and the name is on the tag, so beanie babies with fake tags are flooding into the United States. Collectors have paid a fortune for a "Grunt the boar" made in Shanghai, not Chicago. The fakes are sold over the Internet too, so people have paid a premium for an image of a copy of a stuffed toy.

That may sound insane, but … we are good consumers, after all, and once our basic needs are met, what else are we going to buy but things on which we place an arbitrary value, making a market for goods or ideas, then acting as if they matter?

Living near the home of the Green Bay Packers (a professional football team), it's easy to see how sports and the digital simulation of fantasy games is one way to channel society's free-floating anxiety and hostility into an arbitrary activity. When entertainment (including professional sports) and information services generate the energy flow of a global economy, it is downright unpatriotic not to participate in the madness.

As one paid to make speeches, I see the speaking "industry" in the light of bogus beanie babies. Many speakers spend hours perfecting the presentation of someone else's ideas. They test the winds of popular interest and jump from horse to horse as they ride, following the wind. Of course, we all feed on each other's ideas, but unless we know the difference between snatching food and planting gardens, we live lives of debasing self-deception, drinking from a dribble glass and wondering why we're always wet.

What IS the real value of what we write, or say, or communicate through electronic media? Richard Dawkins popularized the notion of memes, contagious ideas that replicate rapidly. Some memes are life-giving and empowering, but how do we know which ones? In a world in which anyone can say anything, how do we know which beanie baby is real?

The answer, of course, is that no beanie baby is worth a thousand dollars, and only a willed belief in arbitrary value makes us think it is.

The creative ideas that transform our lives are always free.

One of my talks is called "The Stock Market, UFOs, and Religious Experience." In each of those domains, the menu if often sold as if it's the meal. In a world in which money managers take a cut of the total assets they invest regardless of performance, yet 87% failed to beat the averages last year, selling beliefs, images of God, and stories of communication from the mothership ought to be a snap.

The bottom line in all three is that there IS something real eliciting our projections, but the buyer had better learn to separate iron filings from a magnet.

In a knowledge economy, information is capital, but wisdom is gold. And gold is currently devalued.

Powerful ideas are rare, and those who see truths emerging over the horizon like the first light, long before the dawn is even a possibility in most minds, wade against the tide. In an information economy, the new business model is: ubiquity => mind share => brand equity => success. In the realm of ideas and values, we also begin by giving our best away for free. Give, give, give! and the coalescing of positive energy becomes a spiral like stars in a Van Gogh painting, enabling us to live on the energy that eddies back into our lives. How much will there be? Enough—there is always enough, enough to sustain us and remind us what matters most.

Springs of water that originate as trickles swell into life-giving rivers. Even in a world in which free spring-water is bottled and sold.

Which ideas matter most? Those that we can seize and make our own, creative insights or abstractions the apprehension of which immediately changes our lives forever. Truths that set us free. Ideas that like good mentors or good coaches quietly put the reins of our lives back into our own hands.

We can't stop bottom-feeders from stealing ideas, nor software pirates from making CDs, nor thieves from faking beanie babies. But so what? If we understand the real value of ideas in the first place, anybody who disseminates them is doing ourselves—and the world—a favor. Powerful life-giving ideas always diminish the need of our egos for a security, a permanence, a safe harbor that in this world simply does not exist. They transform the very context of our lives, that deep place from which we come to everything else, they create a platform inside our selves or souls on which we stand in the middle of the air and darkness through which we are always plunging. They give us the ability to live from the inside out, grounded in the source of ideas and capable therefore of always generating new ones.

Those real ideas enable us to grasp the patterns that point beyond mere data toward the energy that fills the universe and points in turn beyond itself to that which we do not know how to name much less invent or bottle and sell.

Whistleblowers and Team Players

It was only after whistleblowers came out of the closet during the Great Deflation that *Time Magazine* honored the practice of what team players call "ratting out your pals." Conservative magazines like Time may give lip service to whistleblowing in the abstract but never champion whistle blowers until after they have sung. Instead they support the conditions and practices which make whistleblowers a threat in the first place.

Whistleblowers are a reminder that ethics must be embodied in real flesh-and-blood human beings who put themselves on the line. Unless our deeper beliefs and values become flesh, they are words words words designed to make us feel better, rationalize misdeeds, and send distracting pangs of conscience straight into space.

If you have never known a real flesh-and-blood whistleblower, see the film "The Insider" for a good portrait. The film confirms the conclusion of a Washington law firm specializing in whistleblower cases that lists motivations for whistleblowing—money, anger and resentment, revenge, justice—and eliminates all but one as sufficient to carry a whistleblower through the abuse they will face. Only acting from a pained conscience will sustain a whistleblower through the ordeal.

During a recent speech for accountants about ethics, our Q&A moved quickly into the gray areas where accountants spend much of their time. Outsiders think accountants live in a black and white grid with simple answers but in fact they wade through a swamp of maybe this or maybe that.

Accountants are paid whistleblowers. Accountants are intended to be in the corporate culture but not of it, to use company books like mirrors to reveal the truth and consequences of choices. That's why it is so difficult to do the job right.

The tension comes from the fact that only an individual can have a conscience. An institution or organization can develop a culture that supports doing the right thing only when a leader pursues that objective with single-minded intensity. Left to themselves, all cultures are based on survival, not telling the truth. Cultures reward team players, not whistleblowers. In all my years as a teacher, priest, speaker and consultant, I have never seen a culture with a conscience.

A cop friend reminds me that the first time a rookie cop sees his partners beat someone up in an alley or notices that money or cocaine doesn't always get back to the station, he is closely watched. The word goes out quickly that "he's OK" or "watch out for him." Those that are OK move up. The cop is a practicing Roman Catholic and noted that recent scandals in the church are symptoms of the same dynamics.

Institutions usually encourage disclosure only when it no longer matters. Operation Northwoods—the desire by the Joint Chiefs of Staff in 1962 to eliminate Fidel Castro by sinking refugee boats from Cuba, attacking our own base at Guantanamo, and planting terror bombs in American cities—was revealed by James Bamford in his book *Body of Secrets*, but nary a peep of outrage greeted revelation of the treasonous scheme. When the Church apologized to Galileo for torturing him four hundred years after the fact, it raised the question of how an institution had so lost its moorings that someone might think an absurd gesture like that had meaning.

In Wisconsin a friend was nominated to head an arts board at the state level. His work on behalf of the party in power and his passion for art collecting made him a natural but he was passed over. I asked a confidante of then-governor Tommy Thompson why.

"He's not a team player," he said. "He isn't predictable."

The guy who told me this was a team player. He was faithful and steady and worked tirelessly to raise money for the party. When friends were "naughty," as he called it, he looked the other way. He called recently to tell me he was now a million dollars richer, having been compensated at that level for three years on the board of an energy firm. He had been recommended for the position by his friend, now-Secretary Tommy Thompson.

Thus has it always been. Thus will it ever be.

Why are so many of your heroes, I was asked, people who were assassinated? Why do names like Jesus, Lincoln, Gandhi, and Martin Luther King, Jr. keep showing up in your conversation?

I think it's because they embody what it takes to make a stand on behalf of the truth. They were all human but found the courage to blow the whistle on the cultures of death our institutions create. Their reward was getting whacked.

Make no mistake, those who articulate or embody an upward call always inspire ambivalence. A disciple of Gandhi said that even those who loved him most were secretly relieved when he was murdered because for the moment the pressure was off. Jesus as icon is malleable in the hands of his institutional custodians whereas Jesus the Jew in the street was a real pain.

In an era characterized by increasing secrecy by the government and the gradual but progressive surrender of our rights, it's only a matter of time until some malevolent design ripens and bursts into the sunlight because some whistleblower just can't stand it another minute. Some team player, their motives mixed but their conscience pricked, will tell the truth. That's the only way to have accountability when those with power and privilege remove transparency from the processes of government and business.

When a mainstream Midwest woman asks how she will tell her grandchildren what America was like before the Great Change, how she will explain openness and disclosure, the Freedom of Information Act, guarantees in the Bill of Rights ... then I know that we don't need a weatherman to know the direction of the wind and see the firestorm on the horizon. Signs of the times grow on trees like low-hanging fruit, ripe for the picking.

We are all team players, all of us some of the time, some of us all of the time, but we each have our own particular crossroads where we must decide if our words will become flesh. It is never easy and there are always consequences. Only integrity will see us through to the bitter end and none of us really know if we have it until it is tested.

The Crazy Lady on the Treadmill

We've all had the experience by now. Someone next to us—in this case, the lady on the treadmill at the fitness center—suddenly starts laughing. She didn't snicker as if she had just thought of something funny. No. She laughed loudly for a long long time.

Laughter is a social event. When someone is laughing we want to laugh too. Laugh and the world laughs with you, as they say. I turned without thinking to see what was so funny.

Of course, I couldn't tell because she was inside her own bubble. She was wearing headphones and staring at the small television screen mounted on the treadmill, striding in one of those Monty Python power walks, going nowhere fast and staring at the miniature people on the screen.

The first time it happened was in a coffee shop at the condiment counter. I was stirring my caramel non-fat latte and getting ready to lick the foam from the wooden stick when the woman beside me suddenly began talking. She spoke in a conversational tone except there was no one there but herself. I assumed for a moment she was like the people we pass on the street engaged in animated conversations not exactly with themselves but with ghosts, imaginary companions or antagonists inside the bubble of their heads. Then I saw the spongy ears of the barely noticeable headset and the tiny mic that indicated she was having a telephone conversation.

I like to look at the eyes now of people inside those bubbles and see where they're fixed. In the case of the coffee shop conversation, they were aimed at a point about half way to the vanishing point. I am sure some social scientist is writing a thesis even as we speak—excuse me, even as I type—about the reconstruction of the ratios of social space we inhabit as a result of these bubbles. I know people are saying how irritating it is when someone in a restaurant speaks too loudly on their cell phone because the edges of the social space presumed to be part of our table have been broken and allow that intrusive conversation to cross the line.

I found myself wondering how different it was, really, having one of those electronically enhanced conversations and carrying on with people who aren't there. The blurred margins of the two kinds of experience make a cellphone conversation next to us feel strange because we thought that proximity meant we inhabited the same social space. It's like we're still living on a two-dimensional grid which is suddenly intersected by a portal or node in the abstract four-dimensional spacetime of electronic communication.

Maybe the difference is that someone out there in televisionland as we used to call cyberspace or in a wireless conversation is creating a virtual world self-consciously whereas the ranter is caught in a loop like a stuck record.

But I don't know about that. I often compose things I am writing or practice speeches in my head as I walk in my quiet neighborhood and I find myself gesturing, saying words aloud like someone in headphones singing along. I lose myself in the process and am pulled abruptly from that "rain man space" when I realize that someone is looking at me the way I looked at the crazy lady on the treadmill. I am not really talking to myself but to an audience I am recreating virtually, the one I imagine out there, as I am in this moment as I write. On radio programs I do that too, sitting in a studio with only a producer or in my office doing an interview over the telephone; I imagine a listener and speak to them directly.

Appearances can be deceiving. Back in my preaching days, I recall getting ready for an early church service when a church member whispered conspiratorially that someone had come in that I hadn't met. "You need to connect with her," she said. "She's incredibly rich and often gives large gifts to projects that catch her attention."

I went out to see who she meant. There was a new couple and a new single. The couple were well dressed and greeted me with warm smiles. The solo wore a stained sweatshirt, her dirty hair went in all directions,. and she looked around with a distracted rhythm that made it difficult to connect. You know, of course, that I guessed wrong, that the one who looked like a wandering homeless was in fact the one who was filthy rich. I realized then that the only difference between a street person and an eccentric was a trust fund.

Insanity like wisdom is contextual. Someone said that the difference between a hallucination and a vision is that no one shares the hallucination. So what then is this dialogue between us, a portal in a new kind of spacetime or a closed loop back into myself? If our hallucination is consensual, as Gibson defined cyberspace, where is the emphasis, on the hallucination or the fact that it is consensual? Communities, societies and I guess planetary civilizations can all go crazy but no one knows it inside the bubble unless someone external to the madness says so. Maybe everything is inside a bubble, one we create by pretending to have a conversation. Or maybe the boundaries between us don't really exist except as abstractions like broken lines indicating states on a map. Maybe we're all cells in a single brain having a single dream, maybe the entire universe is like that, the universe itself becoming conscious through the various apertures which we call evolving organisms, each finding itself looking at the rest in a different way and trying to say what they think they see. Maybe the universe is just a crazy lady on a treadmill, going nowhere fast in a wild Monty Python walk, having one hell of a good time, laughing like a madman.

The Enemy is ... WHO?

There are days I miss the Cold War a lot.

Things were so much simpler then. The world was divided into two great camps, ours and theirs, and everybody who didn't fit neatly into the schema could be made to fit with a shoehorn of twisted cold-war logic. Countries irrelevant to the ideological battle were either "with us" or "against us."

One day I looked at the map and realized that—even at the height of the Cold War—the geopolitical domination of the world by the United States was nearly absolute. The enemy controlled much less territory than the USA, but the more the United States controlled, the less secure Americans felt.

I wondered how much we needed in order to feel secure.

That reminded me of a seminar I attended. A man of obvious means was standing up and boasting of making his first million and working on his second. After listening for a long time, the seminar leader asked: How much will you have to have in order to have enough?

The guy was stopped in his tracks. He had obviously never considered the question. He sat back down and went away for a week to think it over.

The next week he stood up again but his manner was different. Finding the answer had changed the way he spoke.

"I realized how much I need in order to have enough," he said. "I have to have ... all of it."

The Cold War worked because it was written in Big Letters. The economic and political struggle was merely a foundation for the cosmic struggle. The Cold War was Armageddon, Good versus Evil.

Then everything fell apart. Our souls are working overtime to find templates onto which to project the evil that lurks in the hearts of all of us. We're not having much luck, but a couple of candidates are running hard—computer hackers and technology itself. Rational people—educated, thoughtful people—suddenly launch into tirades against "what computers are doing to the world" or how "hackers want to break into my computer."

I'll save "computers" for another time and focus on hackers.

Hackers are often portrayed as whacked-out loners hunched over glowing monitors in the night, breaking into our bank accounts.

Those aren't the hackers I know. The men and women at DefCon V included some of the best and the brightest. Many wore the costumes of hacking culture but we know better than to stop there, don't we? Ask what they do for a living and you'll find ranking technocrats from Microsoft, IBM, banks and large consulting operations, makers of the best firewall and security systems. You'll find members of every intelligence agency around. Mingling with the hackers and exchanging information at DefCon and the Black Hat Briefings in Las Vegas were people from the military, CIA, FBI, NSA. In fact, those mingling with the hackers ... were the hackers.

Naturally.

Where else could agents of military and business intelligence learn their craft and subtle art but in the sleepless, passionate quest for knowledge in the networked world, tunneling through the gerbil tubes of the wired world? Hackers are need-to-know machines who go where they need to go to learn what they need to learn in order to understand how things really work.

In the knowledge economy, the people who know how to find knowledge and link it into meaningful patterns, not just amass collections of data but connect it in ways that illuminate and disclose the human reality behind the information—those are the people with their hands on the throttles of power, and those are hackers at their best. Even when they work for multinational corporations and intelligence agencies.

The revolution in information technology is one reason the boundaries of nation states are growing more and more permeable. Those boundaries evolved to define economic and political reality and protect populations. These days, when we can live anywhere and work everywhere, when multi-national corporations make decisions that transcend traditional political structures, when ideological and religious passions surge back and forth over borders like the ocean at high tide—how are we to identify the enemy?

According to the Chinese Army newspaper, Jiefangjun Bao, speeches at the new Military Strategies Research Center were summarized this way:

The goal is no longer to preserve oneself and destroy the enemy. The goal is to preserve oneself and control one's opponent. (*Wired*, May 1997).

Information warfare is the name of the game.

Remember Somalia? All it took was a thirty-second video clip showing one Marine being dragged through a crowd to undermine our will.

I miss those Cold War spy novels. Double agents unmasked at last, spymasters playing global chess. Today the game is more like ten-dimensional chess. Allegiance to who and to what? Our loyalties are nested like Russian dolls, and sometimes even we don't know on behalf of whom or what we are really acting.

Behaviors that are penalized in one context are sanctioned in others. The determining factor is not the action but the allegiance of the actor.

It is not the behavior of hackers that is threatening but their perceived allegiance. Ally themselves with a government, intelligence agency, or large corporation and they can hack their hearts out, with more computing power at their disposal than they can dream.

Ironic, isn't it? The "West" won the Cold War, the economy is the best in decades, and Americans have everything they thought they wanted. Yet people are more anxious and insecure than ever.

Security comes not from having what you want—even when you have all of it—but from stability and predictability. And that—in a world undergoing fundamental transformation—is in short supply.

So people will continue to project fear and anxiety onto templates provided by the media. Hackers—not criminal hackers, but real hackers—will continue to be demonized and misunderstood, because the nature of power, influence, and leverage in a global knowledge economy will continue to be misunderstood.

You can't write a 32-bit application for an IBM XT. It just can't handle the code.

I won't remind you that Pogo said "We have met the enemy and it is us" because everybody knows that. I'll just note that no amount of stuff, including knowledge, can give us what we need. Our desperate search for security in a changing world is really the pursuit of our own souls. That's all we can ever possess in its entirety anyway. We're like people wearing glasses running around frantically looking for our glasses. We have what we need, always, here and now. The enemy is anybody and anything that prevents us from seeing clearly what a crazy maze we have built to keep us from remembering that.

Life in Space

There was so much hullabaloo at Def Con VI, (the recent convention for computer hackers, journalists, screen writers, producers, computer security and insecurity experts, programmers, federal agents, local police and sheriff's deputies, advertisers and marketers, hotel security guards, undercover agents, refugees from raves, groupies, and endlessly curious mind-hungry men and women of all sorts and conditions)—hullabaloo, that is, about how hackers have morphed from evil geniuses into respectable men and women operating at the highest levels of industry and commerce, the military, and the intelligence community.

The basis for comparison, of course, is an image of hackers as whacked-out loners hunched over glowing monitors late into the night, cackling like Beavis or Butthead as they break into our bank accounts—an image created and sustained by the media.

Well ... let's be real. Some do, some are. That's part of the scene, the digital equivalent of growing up in Hell's Kitchen and living down these mean digital streets. That, however, is not the essence of hacking.

Hacking is curiosity, playfulness, problem-solving, motivated by the pleasure of browsing, following one's nose where others say it doesn't belong, looking for a constellation in the seemingly random stars. Following the luminous bread crumbs deep into the twilight forest. Building an elusive, always-hypothetical whole that forms and dissolves and forms again at every level of the fractal puzzle of life.

Hacking has its roots in Renaissance men like da Vinci and Machiavelli who saw clearly and said what they saw.

But something else is happening too. As I looked out at the audience of the Black Hat Briefings, I saw that the roles of journalists, specialists in competitive intelligence, spies, even professional speakers like myself, were converging, that roles in a digital world are as fluid as identities.

The skills of hackers and intelligence agents are the skills needed in the virtualized worlds we are learning to inhabit.

We hear endlessly of convergence of form and structure in the wired world. Every digital interface is an arbitrary distinction. Because we can reconstitute bits in whatever form we like, deciding to call an interface a PC, TV, or PDA is a job for marketing, not engineers.

But I'm talking about the convergence of roles. The digital world is engineering us in its image. Because that world is interactive, modular, and fluid, our lives are too. We don't even notice anymore that to choose to present ourselves to the world is a choice.

At one extreme, identity hacking—stealing identifiers like numbers and codes with which to gain access to the social and economic world or creating a new identity from whole digital cloth in order to disappear and surface in a new body—is a growing industry. But choices we take for granted—changing jobs, religions, marital status, changing our names, changing careers, changing who we essentially think we are—have become part of consensus reality. Not so long ago, people who did that were thought to be just plain nuts.

Once upon a time, the roles we were expected to fulfill were our destinies. Unless external crises intervened, people were expected to stay in one place, get a job and keep it, get married and stay married, be whatever religion they were told they were (as if something else were even thinkable), and live inside a single identity that was so much a fish in water that it wasn't questioned.

Identity is a social construction of reality that's noticed only when the external factors that shape it have changed.

The new consensus reality is reinforced by information sources from talk shows to the Wall Street Journal. We can choose careers, another marriage, another religion, another way of being ourselves, and we are everywhere surrounded by helpful advice about how to do it.

In the digital world, sanity means having the resources and capacity to know how to morph, changing presentations that are bridges between constantly shifting external factors and our own developmental stages. This is true for organizations as well as individuals.

The protean self, back-engineered from the structures of our information technologies, thinks of life as a creative act. The ability to distinguish who we are from our presentations, knowing how to use those presentations to exercise power, build feedback loops of energy and information to sustain us, that's a skill that used to belong to spies alone. Now it's asked of everyone who wants to remain viable.

Hackers call it social engineering, learning how to look and sound a particular way to elicit the information needed to build the Big Picture. In business, it is often called competitive intelligence. Some just call it "the way it is."

Every time I say, "the edge is the new center," I notice that the edge I had in mind is no longer the edge. A new edge is emerging. Turn-around time is about six months, not only for computers, but for viable constructions of reality.

We work and live in space stations, docking in modular fashion, then we're off again into space. That space is sheer possibility in which we create literally from nothing. The pull of the future creates the irresistible shapes of present possibilities with which we must comply. Every time we break through to a new way of seeing things it feels momentous, but breakthroughs are momentous for only a moment. Then they become commonplace, the background noise of the next stage of our lives.

Evil genius hackers? Give me a break. The hackers who have their hands on the throttles of power in the digital world were "kids" three years ago. That's about as long as a current generation lasts. And civilization too is ramping up toward a single point of convergence where identities are arbitrary. What we call "our species" will soon be a wistful memory in the molecular clusters of the progeny we design, an arbitrary distinction that served us for a while before we morphed. A noun turned into a transitory verb. Ice turned into a flowing river.

Hacking Chinatown

"Hacking Chinatown" was written for *South Africa Computer Magazine*, then republished by the *Computer Underground Digest* and *CTHEORY*.

"Forget it, Jake. It's Chinatown."

Those are the last words of the movie *Chinatown*, just before the police lieutenant shouts orders to the crowd to clear the streets so the body of an innocent woman, murdered by the Los Angeles police, can be removed.

Chinatown, with Jack Nicholson as Jake Gittes, is a fine film: it defines an era (the thirties in the United States) and a genre—film noir—that is a unique way to frame reality.

Film noir is a vision of a world corrupt to the core in which nevertheless it is still possible, as author Raymond Chandler said of the heroes of the best detective novels, to be "a man of honor. Down these mean streets a man must go who is not himself mean, who is neither tarnished nor afraid."

Chinatown also defines life in the virtual world—that consensual hallucination we have come to call "cyberspace." The virtual world is a simulation of the "real world." The "real world" too is a symbolic construction, a set of nested structures that—as we peel them away in the course of our lives—reveals more and more complexity and ambiguity.

The real world IS Chinatown, and computer hackers—properly understood—know this better than anyone.

There are several themes in *Chinatown*.

(1) People in power are in seamless collusion. They take care of one another. They don't always play fair. And sooner or later, we discover that "we" are "they."

A veteran police detective told me this about people in power.

"There's one thing they all fear—politicians, industrialists, corporate executives—and that's exposure. They simply do not want anyone to look too closely or shine too bright a light on their activities."

I grew up in Chicago, Illinois, known for its political machine and cash-on-the-counter way of doing business. I earned money for my education working with the powerful Daley political machine. In exchange for patronage jobs—supervising playgrounds, hauling garbage—I worked with a precinct captain and alderman. My job was to do what I was told.

I paid attention to how people behaved in the real world. I learned that nothing is simple, that people act instinctively out of self-interest, and that nobody competes in the arena of real life with clean hands.

I remember sitting in a restaurant in a seedy neighborhood in Chicago, listening to a conversation in the next booth. Two dubious characters were upset that a mutual friend faced a long prison term. They looked and sounded different than the "respectable" people with whom I had grown up in an affluent part of town.

As I grew up, however, I learned how my friends' fathers really made money. Many of their activities were disclosed in the newspaper. They distributed pornography before it was legal, manufactured and sold illegal gambling equipment, distributed vending machines and juke boxes to bars that had to take them or face the consequences. I learned that a real estate tycoon had been a bootlegger during prohibition, and the brother of the man in the penthouse upstairs had died in Miami Beach in a hail of bullets.

For me, it was an awakening: I saw that the members of the power structures in the city—business, government, the religious hierarchy, and the syndicate or mafia—were indistinguishable, a partnership that of necessity included everyone who wanted to do business. Conscious or unconscious, collusion was the price of the ticket that got you into the stadium; whether players on the field or spectators in the stands, we were all players, one way or another.

Chicago is Chinatown, and Chinatown is the world. There is no moral high ground. We all wear masks, but under that mask is ... Chinatown.

(2) You never really know what's going on in Chinatown.

The police in Chinatown, according to Jake Gittes, were told to do "as little as possible" because things that happened on the street were the visible consequences of strings pulled behind the scenes. If you looked too often behind the curtain—as Gittes did—you were taught a painful lesson.

We often don't understand what we're looking at on the Internet. As one hacker recently emailed in response to someone's fears of a virus that did not and could not exist, "No information on the World Wide Web is any good unless you can either verify it yourself or it's backed up by an authority you trust."

The same is true in life.

Disinformation in the virtual world is an art. After an article I wrote for an English magazine about detective work on the Internet appeared, I received a call from a global PR firm in London. They asked if I wanted to conduct "brand defense" for them on the World Wide Web.

What is brand defense?

If one of our clients is attacked, they explained, their Internet squad goes into action. "Sleepers" (spies inserted into a community and told to wait until they receive orders) in usenet groups and listservs create distractions, invent controversies; web sites (on both sides of the question) go into high gear, using splashy graphics and clever text to distort the conversation. Persons working for the client pretend to be disinterested so they can spread propaganda.

It reminded me of the time my Democratic Party precinct captain asked if I wanted to be a precinct captain.

Are you retiring? I asked.

Of course not! he laughed. You'd be the Republican precinct captain. Then we'd have all our bases covered.

The illusions of cyberspace are seductive. Every keystroke leaves a luminous track in the melting snow that can be seen with the equivalent of night vision goggles.

Hacking means tracking—and counter-tracking—and covering your tracks—in the virtual world. Hacking means knowing how to follow the flow of electrons to its source and understand on every level of abstraction—from source code to switches and routers to high level words and images—what is really happening.

Hackers are unwilling to do as little as possible. Hackers are need-to-know machines driven by a passion to connect disparate data into meaningful patterns. Hackers are the online detectives of the virtual world.

You don't get to be a hacker overnight.

The devil is in the details. Real hackers get good by endless trial and error, failing into success again and again. Thomas Alva Edison, inventor of the electric light, invented a hundred filaments that didn't work before he found one that did. He knew that every failure eliminated a possibility and brought him closer to his goal.

Listen to "Rogue Agent" set someone straight on an Internet mailing list:

"You want to create hackers? Don't tell them how to do this or that. Show them how to discover it for themselves. Those who have the innate drive will dive in and learn by trial and error. Those who don't, comfortable to stay within the bounds of their safe little lives, fall by the wayside.

"There's no knowledge so sweet as that which you've discovered on your own."

In *Chinatown*, an unsavory character tries to stop Jake Gittes from prying by cutting his nose. He reminds Gittes that "curiosity killed the cat."

Isn't it ironic that curiosity, the defining characteristic of an intelligent organism exploring its environment, has been prohibited by folk wisdom everywhere?

The endless curiosity of hackers is regulated by a higher code that may not even have a name but which defines the human spirit at its best. The Hacker's Code is an affirmation of life itself, life that wants to know, and grow, and extend itself throughout the "space" of the universe. The hackers' refusal to accept conventional wisdom and boundaries is a way to align his energies with the life-giving passion of heretics everywhere. And these days, that's what needed to survive.

Robert Galvin, the patriarch of Motorola, maker of cell-phones and semi-conductors, says that "every significant decision that changes the direction of a company is a minority decision. Whatever is the intuitive presumption—where everyone agrees, 'Yeah, that's right'—will almost surely be wrong."

Motorola succeeded by fostering an environment in which creativity thrives. The company has institutionalized an openness to heresy because they know that wisdom is always arriving at the edge of things, on the horizons of our lives, and when it first shows up—like a comet on the distant edges of the solar system—it is faint and seen by only a few. But those few know where to look.

Allen Hynek, an astronomer connected with the U. S. Air Force investigation of UFOs, was struck by the "strangeness" of UFO reports, the cognitive dissonance that characterizes experiences that don't fit our orthodox belief systems. He pointed out that all the old photographic plates in astronomical observatories had images of Pluto on them, but until Clyde Tombaugh discovered Pluto and said where it was, no one saw it because they didn't know where to look.

The best computer consultants live on the creative edge of things. They are pathfinders, guides for those whom have always lived at the orthodox center but who find today that the center is constantly shifting, mandating that they learn new behaviors, new skills in order to be effective. In order to live on the edge.

The edge is the new center. The center of a web is wherever we are.

When I looked out over the audience at DefCon IV, the hackers' convention, I saw an assembly of the most brilliant and most unusual people I had ever seen in one room. It was exhilarating. We all felt as if we had come home. There in that room for a few hours or a few days, we did not have to explain anything. We knew who we were and what drove us in our different ways to want to connect the dots of data into meaningful patterns.

We know we build on quicksand, but building is too much fun to give up. We know we leave tracks, but going is so much more energizing than staying home. We know that curiosity can get your nose slit, but then we'll invent new ways to smell.

Computer programmers write software applications that are doomed to be as obsolete as wire recordings. The infrastructures built by our engineers are equally doomed. Whether a virtual world of digital bits or a physical world of concrete and steel, our civilization is a Big Toy that we build and use up at the same time. The fun of the game is to know that it is a game, and winning is identical with our willingness to play.

To say that when we engage with one another in cyberspace we are "Hacking Chinatown" is a way to say that asking questions is more important than finding answers. We do not expect to find final answers. But the questions must be asked. We refuse to do as little as possible because we want to KNOW.

Asking questions is how human beings create opportunities for dignity and self-transcendence; asking questions is how we are preparing ourselves to leave this island earth and enter into a trans-galactic web of life more diverse and alien than anything we have encountered.

Asking questions that uncover the truth is our way of refusing to consent to illusions and delusions, our way of insisting that we can do it better if we stay up later, collaborate with each other in networks with no names, and lose ourselves in the quest for knowledge and self-mastery.

This is how proud, lonely men and women, illuminated in the darkness by their glowing monitors, become heroes in their own dramas as they wander the twisting streets of cyberspace and their own lives.

Even in Chinatown, Jake. Even in Chinatown.

Mutuality, Feedback, and Accountability

"Coping with exponential change means using traditional spiritual tools ..."

If we're lucky ... after we have exhausted every other avenue ... we turn to the traditional tools of spirituality to cope with stress and rapid change. That is, with life on earth.

You won't hear a word like "spirituality" in many Fortune 500 board rooms, but businesses too are figuring it out.

It makes sense. These practices are the result of centuries of reflection on our collective experience, sifting and sorting everything we've tried in order to find out what really works.

Some things seem to work across the board.

Effective organizations have: (1) a high degree of mutuality; (2) mechanisms for frequent, widely distributed feedback; and (3) accountability to mutually-agreed-upon goals and to a vision of leadership. I have said that before but let me spell it out.

MUTUALITY:

The individualism that many of us were taught was axiomatic to being human was in fact generated by a print culture. Before the Gutenberg era, nobody thought that way.

Digital culture undermines individualism and our ability to act as if we exist apart from our communities.

A representative of a school district told me they received good grades from corporations that hired their graduates, except in one area: cooperative learning. I asked for a definition of "cooperative learning" and realized that in my day it was called "cheating."

This indicates how completely the assumptions of post-WW2 America have been turned upsidedown by the digital revolution.

Independent learning apart from the learning of the group or organization is not viable because the individual—one of many modular units in the network—MUST correlate with the activity and trajectory of the group.

This means learning new behaviors.

A CEO of a utility told me they used to spend 85% of their time on task and 15% on process issues—leadership, team work, and the like. Today the percentages are reversed, not because utility executives want hugs on Monday morning but because that's what mandated by conditions.

As a young hacker said, looking back nostalgically on the days when he could know everything about hacking a system: "These days, there's too much for any one person to know. That's why the most important thing I need to know is what I don't need to know. The second most important thing is, who knows it? So I can get it when I need it."

He needs to know ... people-skills, networking, cooperative learning. The electronic network has back-engineered a culture that must work as the network does—in order to be effective AND in order to function in a symbiotic relationship with the network.

FEEDBACK:

An organism in a rapidly changing environment must know what is happening inside itself (the smaller system) and outside. As we climb a spiral toward more and more complex organization, the feedback loops must grow larger, and those inside the boundaries of the smaller system must make us aware of internal changes in equilibrium. That's why quality programs insisted the system must include suppliers, customers, and ultimately all "stake-holders"—to remain current with conditions and to capture the knowledge of everyone relevant to the success of the organization.

All the way up the spiral, every living organism is part of the system and ultimately the universe must factor in the input from every living creature. That's another way of saying what mystics have always said, that all life is interdependent and the energy and information that constitutes the visible interaction among all parts of a system are an image of a universe that is unified, self-conscious and evolving.

On a more mundane level, we can not adapt to changing conditions or "morph" unless our sensors are out there in the multiplicity of environments that impact our well-being.

And ... we must hold ourselves accountable for what we learn.

ACCOUNTABILITY:

When things are going well, accountability diminishes. Then when things don't go well, there's chaos.

Many financial managers are riding the crest of a remarkable bull market. They receive handsome fees to manage billions of dollars. Yet only 7% of professional fund managers beat the S&P index over the past year.

So long as benchmarks are not used to align performance with stated goals, those managers can continue to benefit from the rising tide that is lifting all boats. But when the turnaround comes, mutual funds by the hundreds will merge or collapse in a shake-out that imposes stricter accountability.

Without accountability, there is no way to look back and see how we have done and no way to look forward to see where we want to go.

The absence of any of these essential qualities skews an organization in predictable ways.

Mutuality and feedback without accountability result in team-work and data-exchange to no end. The system has no compass and no means of realigning itself when it gets off course.

Feedback and accountability without mutuality characterize a top-down system driven by goals. But the feedback becomes fragmented, undermining the security we need to function. An anxious and fearful system becomes rigid and isolated; feedback does not result in purposeful action.

Accountability and mutuality without feedback characterize a closed system that is blind. Cults work well for those inside them until they don't. When something new gets in—an idea, a fact, an event—it's a debilitating blow. The cult either commits suicide or is transformed into something else. (A business can be a cult, too.)

To surrender one's illusion of individual well-being in order to participate in a larger structure for the good of all (including oneself) is, paradoxically, an individual decision.

It is a moment of insight into one's real nature and destiny that so threatens our illusions and habitual state of denial that it has been called "a nightmare in daylight." Once it happens, however, we can never again think of ourselves as we did before. When the paradigm changes, there is no going back.

This template is not a recipe. It simply identifies some of the marks of that perilous journey, undertaken with fear and trembling appropriate to the real risks and real rewards.

The Day the Computer Prayed

When a computer prays, is it really prayer?

And I mean real prayer, I don't mean some mood-altering self-manipulation. I mean, is there an intentional focus of energy and intelligence, the intelligence of the heart, so that something happens beyond the merely subjective, something that percolates powerfully through all the levels of our consciousness?

Nor do I mean merely ritualistic prayer. As T. S. Eliot said,

> "... prayer is more
> than an order of words, the conscious occupation
> Of the praying mind, or the sound of the voice praying."

Not that anything's wrong with that. We human beings need repetitive symbolic acts to comfort or sustain us. No matter how spontaneous we think we are, our habits dig ruts in our psyches, and the wagons of our lives—including prayer—roll in those ruts.

That's at the top level, that's what happens when a conscious mind thinks or says words. Which means that one assumption of prayer is that telepathy is real.

A clairvoyant moment returning from the moon became a new axis for Edgar Mitchell's life. The astronaut experienced a silent communication that disclosed to him the unity of all things. It was a moment of transcendent communion that told him he was right where he belonged, at home in a gregarious universe.

When enough people have that kind of experience, they come together and struggle to articulate what they believe. Our individuality is expressed fully in community. That makes for corporate rituals. Our religious institutions and organizations are grievously flawed, but they do bridge the generations and pass on a legacy of the symbols of possibility and promise that can then explode once again into our real experience.

Or devolve once again into ritualized prayer.

Monasteries sometimes send notices saying they are praying for me on a particular day. The names for whom they pray are in a book or rolodex, and as the wheel of time turns, so does the rota. Names are plugged automatically into the blank spaces.

It's a natural for computerization. That monastery, executing its structured top-down program for ritualized prayer, is a symbol manipulating machine. All human beings are symbol manipulating machines, and so are computers. We're interlocked in a symbiotic embrace that is taking both networked computers and networked human beings up a spiral of mutual transformation. As we are changed by this process, the symbols we use to express our understanding of the process also change.

Whenever there is a transition from one "technology of the Word" to another, there is always resistance. In an oral community, prayer and ritual were alive as the words were uttered. When the words were written down, they seemed a pale reflection of words that had potency when spoken. Same thing when the printing press with movable type was invented. Some people just couldn't read what was printed.

One of my favorites is the Duke of Urbin. A passionate manuscript collector, he refused to read printed books, but when he heard of one that he wanted, he had it delivered to the monastery where the monks copied it over by hand.

So there will be resistance to prayers on a computer, as if they are somehow "not real." Anyone who has reached out on-line during a crisis, however, as I did recently when my brother was deeply depressed and threatening suicide, knows that the words that show up on the monitor in the middle of the night are words of light and life. The response I received to my late-night invitation to a few colleagues and friends resulted in action that saved my brother's life.

Some of those people prayed and communicated their prayer via email. Some lighted simulated candles in the digital darkness, no less candescent for being words or images. Some typed advice or encouragement. And some, like those monks that turn that rolodex, plugged in our names to programs they had written and let the computer just keep on praying for us, day after day.

OK, you tell me: Did the computer pray?

But remember, the "computer" is not some stand-alone machine, a brain in a bottle on somebody's desk, the computer is the global network, alive not only with energy but also with intentionality.

Intentionality transforms what would otherwise be rote into prayer. Whether people are reading in a chapel or communicating on the Network, the symbols—printed words or luminous pixels—are turned to flame.

Prayer becomes real, according to Hasidic Jews, when the words on the page become flame. The words on the monitor too.

The Network is a symbiosis, a global community of human beings interlocked with a global network of millions of computers. The symbiosis—digital humanity expressing itself through an electronic network inseparable from itself—prayed and keeps on praying.

How can we separate a mediating structure from that which becomes incarnate in and through it? The medium is the message, the context content, and words becomes both silicon and flesh. Physical objects are permeated with memory and meaning, and then they become sacred. Sacred places are spaces bracketed in the physical world—which is nothing but energy also—transformed by the energy and intelligence of the heart.

There are moments—aren't there?—when we feel the presence of someone interacting with us through the monitor, through the modem, through email or IRC, so palpably that we feel them there in the room, we feel their energy, their intention focused on us, and we feel it as well when their focus shifts and the energy wanes. Being is manifest through the wires and electromagnetic energy of the digital world.

So yes, prayer happens, but the initial question—does a computer pray?—dissolves into finer distinctions as our sensory extensions are fused with our will and intelligence. Prayer happens, luminous sacred spaces glow in the night, the sudden candescence of a million monitors transformed into altars and sacred groves, a roar of flame in the darkest hour.

An Owl in Winter: Millennium's End II

Late last night, I was walking alone on a path leading toward a footbridge that crosses a ravine near my home. As I approached the bridge, I expected to hear the familiar echo of boots on wooden planks but instead I heard—in a moment of silence just before I walked onto the bridge—the first owl of the winter.

I looked at the shadowy branches but couldn't see him. Last winter I saw an owl only once. I looked up from the icy snow and saw him perched on a low branch against a full moon. Then he opened wings that expanded unimaginably as he lifted off into the darkness.

That owl is a bird of prey, not some cartoon bird wearing glasses.

My wife recently attended a speaking engagement with me in Des Moines, Iowa. She had not returned to her home town since we buried her mother fifteen months before. We went to the cemetery and looked for her parents' graves. The markers were already overgrown with grass and covered with leaves. I pulled out the grass and brushed away the leaves so she could see their names, feeling like King Canute ordering the tide to recede. The grass will grow back but next time no one will be there to pull it away.

We walked afterward in a park where my wife had ice-skated as a child. It was one of those lingering twilights when we feel deeply the transitoriness of all things, that all sanctuary is a momentary turning from the wind toward a temporary refuge. We followed a path in the woods along the Raccoon River until it grew dark, and when we turned back, we saw against the sky the silhouette of an owl in the low branches.

Dylan Thomas wrote in Fern Hill that all through the summers of his childhood ...

> *"as I rode to sleep the owls were bearing the farm away ...*
> *in the moon that is always rising ..."*

Owls. Rising moons. Images of death but not of death only, images too of life, life lived on a razor's edge, life thrown at us, fired at us point blank from the barrel of a gun.

The millennium's end IS an ending, of sorts: We will never exactly come this way again, or if we do, we'll all be different. An owl in winter signifies the cold darkness surrounding the solace of our conversation, built of illusory pixels and made luminous by the sheer force of will and desire. Eros in its digital essence, binding us together. A fusion of fact and imagination.

I wrote a few weeks ago about Digital Autumn, and a reader said, it is one thing to embrace the ambivalent longings of autumn when love and loss are inextricably intertwined, when the wet earth is redolent of spring and autumn alike. But it's something else to embrace winter when it finally arrives.

How, he asked, do we welcome the real winters of our lives?

I thought of my first winter in England. At that northerly latitude, as the winter solstice neared, the sun rose lower and lower in the sky, looking like a distant pale candle seen through a frosted windowpane. I understood why Celtic legends tell of the twilight of the gods and eternal winter. There was an icicle of fear in our hearts that the sun would not bounce, but would sink lower and lower and finally disappear.

But the sun did bounce. Winter-blossoming cherry trees flowered. Love unexpectedly melted our hearts once again.

There is no yin without a dot of yang, no yang without a dot of yin. Winter and summer, fire and ice. A figure-eight held in paradoxical tension, traced by a skater on thin ice.

Leafing through a friend's fortieth high school reunion book, I noticed that almost everyone said the same thing when asked what made for a wonderful evening. It was always dinner with family or friends, a quiet evening at home, an afternoon off with someone we love. Now it sounds like many of us are planning the same kind of quiet celebration for the last night of the millennium, just as we would if the universe were ending.

There will be parties too, of course, and a few crazies will do what they can to ruin everything, but most of us just want to be with people we love as we begin not so much the next thousand years as the next few minutes of our lives.

We don't live for millennia. We live for Now.

Digital winter is the hollowness we sense outside the warm glow of our ingathered circle. These digital symbols comfort us with intimations of promise and possibility. Sometimes late at night when we can't sleep, we sit in front of our monitors as before a dying fire, hoping for a sign of communion. We build a digital bridge but before we cross it, we hear in the late-night silence the sound of a solitary owl. We realize that loss and grief have built the hearth on which we light this blazing fire, made up of air, nothing but thin air.

The loss of those we have loved has an icy core but contains a trace of consolation, the suggestion of a tide that will rise once more. Winter-blossoming cherry trees flower. A river of fire flows through these wires. We are alive. Here and now, we are alive. The sun and the moon are rising. And the fact of our being transfigures the threat of annihilation into a bonfire celebrating the end of everything and nothing, destroying and creating worlds.

for Karin/Krystalia—a bright young hacker girl who died too young - in loving memory

Night Light

All my life I have explored imaginary landscapes. From short stories to hypertext and interactive fiction, I have been fascinated by the way narratives are knitted together, ensnaring our minds in complex webs of luminous symbols and deceiving us into mistaking those illusory landscapes for reality itself.

Brains filter out as well as filter in. We wear our brains like blinders, protected from the splendor of the universe so we can keep our noses to the daily grindstone. Most of what matters most is beyond our grasp.

The deeper meanings of our lives manifest themselves as intimations that show up not in words but in the silence between words. Maybe that's why Buddhists say, those who know do not speak, those who speak do not know. Our images and symbols can take us to the gates of that nether-domain of darkness but then we must lower ourselves alone down into the caverns … hang like a spider on a thread … and let go of the rope.

And plunge into the cave of unknowing. Into darkness brimming suddenly with radiant light.

The midnight of the millennium has passed. It is still a ticklish world out there, but for the moment we are all breathing a little easier.

I asked myself, if this were my last opportunity to say what I have learned, to leave these words as a legacy for my children, what would I say?

Of that of which we cannot speak, said Wittgenstein, we must be silent. And humility, said Eliot, humility is endless.

But still, I would like to try.

Computer programmers often learn to think in terms of top-level languages like C but nested in those friendlier languages are languages like assembly that describe the operations of computing at a different level of precision. These days, assembly language is often difficult to distinguish from machine language. That makes it hard to say what computers are "really" doing.

Machine language organizes how logical processes operate. But the process of manufacturing logic is too complex for humans alone, so the chips themselves contain many of those processes. Our network is manufactured by a symbiosis of human beings softwired into designs in turn hardwired into millions of invisible switches.

When we try to look under the level of machine language to the flow of electrons or photons, we spin dizzily down into the depths. When we try to look under our lucid explanations of biological processes to the genetic code that generates them—a code to which it seems we must only add water to grow symbol-manipulating machines like ourselves—we grow equally dizzy. Our identities cascade into a recursive process that spirals down into the darkness.

What is the ultimate source of our identities?

Our identities both as individuals and as civilizations seem to emerge from a recursive process acting on an unknowable initial condition. That's a way of saying that our origins are mysterious. "We" have bootstrapped ourselves from that unknowable condition into the light, hauling ourselves up by our own symbols, just as the universe has bootstrapped itself through successive levels of complexity via sentient beings distributed throughout all of unknowable space.

So when we look into that luminous darkness ... who are "we?" If we try to respond to that question, language breaks. All we can say is that "we" are not who "we" thought we were.

We are more than a community or a collective. Sentient life is woven in a singular web shot through with the light of luminous symbols. We are part of one another and of everything. We belong to one another and we belong to the universe.

We are not merely "fulfilled in community"—we are fulfilled when we so yield ourselves that we lose ourselves and our power is transformed into the power of others like ripples amplified endlessly in the still pool of the universe.

Here at top-level, high up at the level of language, all we receive are hints. At an immense distance—even at opposite ends of the universe—when those we love are hurting, somehow they let us know. At the moment of death, they often manifest themselves. Our connections are protected. We feel their presence as certainly as we know when someone has entered a room. And they have, in a way, entered a room, they have etched themselves indelibly in our souls. When we are socketed deeply into the love of others, we experience their energies whenever they extend in our direction and we know it.

Prayer is a name for clairvoyance radiant with omni-directional power. Telepathy means that we "get it" when it happens. Mystics have described this domain with precision, but a mystic is just an ordinary person who said what was so.

It can be startling to find our words, posted long ago to a newsgroup, archived at Deja News. Or old email stored on a server. Or the pattern of our lives linked and mined and sold to a list. So it ought to be even more surprising when we learn for the first time that all of our words and actions have a life of their own, that our lives are feedback loops of actions and reactions in which we live like ghosts in machines, going along for a ride. Yet paradoxically, our deepest intentions have always generated the patterns of our lives. Freedom and necessity fuse in a way that only the language of paradox can say.

So our ultimate task seems to be to try and try again to align our actions with our best intentions. Our lives will reflect our intentions anyway no matter what. Our lives tell us what we really meant if we have the courage to look at what we did.

And once we know, we can never forget.

An inner compass points always toward True North. The universe is a gesture intimating generosity and gregariousness, the primacy of what we call "love." Life is an invitation to discover somehow despite the seeming evidence to the contrary the means of unceasing affirmation.

Of course we love stories! Stories tell stories, after all, that cannot be told. Every technology of the Word—spoken, written, printed, transmitted—makes this real matrix in which we live a little more visible. These digital images that string us like beads or drops of dew on a spider's web are like dye in the arteries of our Soul(s).

The ghost in the machine is no pale shade. It is more than a luminous mist, more than a demented spirit hungering for its home. It is an intelligence transcending nested levels of insight into the darkness in which wings of the spirit infold into a single point of light and almost disappear.

For Julie, Barnaby, Rachel, Aaron, Scot, Susan, and Jeff

Invitation to a Seance

The email message came literally out of nowhere.

"Hi this is new to me is th5s how yuo do it/?"

The return address contained the first name of a friend I had not heard from in years. I'd heard he was married now and living in San Diego. Beyond that—nothing.

Yet here he was again ... I think.

Wasn't he?

I admit I can be a little cynical. I sometimes feel like Jane Wagner, who said, "I'm getting more and more cynical all the time and I still can't keep up." Sometimes that seems like a reasonable response to the things our species does.

A cynic, however, is really a disappointed idealist. A realist, someone who takes life exactly as it is, on its own terms, fired at us at point blank range from the barrel of a gun, is never disappointed. What happens is just "what's so."

A cynic, on the other hand, is always hopeful but frequently disappointed. My head is a cynic but my heart is naive. My heart is always ready to believe in the highest possibilities. An attitude of forbearance, generosity of spirit, and hope for reconciliation may not always be supported by the evidence, but it does help us get some sleep.

What does that have to do with that email?

Hanging out with hackers who really know their stuff has given me a profound appreciation for the illusions of cyberspace. An email can be easily forged - email address, routing information, IP address, everything in the header. Web sites can be altered or simply appropriated, complete with a URL that brings unsuspecting email to your dummy site.

Cyberspace is full of Potemkin villages.

So ... without a context for that email, how did I know who had written it? How could I know?

I wrote back, "Mike??? Is that really you?"

He wrote: "yes i am jst lrnig how to do this i cnt type"

Flashback to when I lived in England, a young man in a village in Surrey near London. I had made the acquaintance of an elderly woman named "Mrs. Frazier." I don't remember her first name.

Mrs. Frazier lived in a nursing home and from time to time, we invited her to our home for "high tea." She loved "getting out of prison," and from my point of view, the visits were always interesting.

Right in the middle of a sentence, as she lifted her teacup to her lips, Mrs. Frazier would suddenly pause and freeze, her eyes fixed, then set down the teacup and begin sputtering, shaking like a large wet dog coming out of a pond. Then she said, "That's all right, friend," to the air just behind her left ear, and to us: "He's here. He's with us."

That meant her "Indian guide" had manifested himself, and for the next while, she passed along tidbits of information about "life on the other side" and insights into our progress at our current "level of vibration."

Sometimes the guide brought messages from the dead. I never believed them, and I always wanted to believe them. They were always encouraging, but seldom specific. Once in a while I asked a test question that required knowledge of past events. The answer was always too vague to be evidential.

So we were left with Mrs. Frazier, a dear lonely woman who loved her outings and "specialness" so much she would let her imagination take over and for a few moments become the center of our attention.

I sent test questions to my email friend. One requested "the name of the mascot of the high school social athletic club that you belonged to, and where did it live."

Mike did better than Mrs. Frazier.

"It ws a chickn," he replied. He added the unprintable nickname we had given the hapless fowl. 'It lved on the roof of the apartment bildng you lvd in untl they found out and made us give it away because of livestock laws."

He passed the test. My old friend really was new to cyberspace and learning (like a ghost after it leaves its body) how to represent himself as a digital construct interfacing with other digital constructs.

He types better now. In one email, he mentioned a woman we both knew and asked where she was. To my astonishment, an email from her arrived—out of nowhere—that week.

"It is amazing," Mike wrote, "that you received an e-mail from her at the very time we were talking about her. Is it possible that e-mail is just a way of communicating with one's self? i.e. it doesn't get sent anywhere except into the brain of the sender?"

I think he's right, but the Self with which we communicate is the larger Self, the overSelf in which we all participate, the single organism constituted by all of our humanity. Of that Self the Internet is a collective representation, the echoes or reflections of our digital dreaming and thinking.

The Net is an imaginary garden with real toads in it. We create it together literally out of nothing, then forget that we made it up so we can play in it.

The eyes with which you read this message are one pair of the multi-faceted eyes of a single honeybee looking at a digital simulation of our hive mind.

This digital transmission can be faked or hijacked, text changed in transit, images scribbled over with graffiti, but I don't think that's the last word.

Connection is the last word, genuine connection, and the community that blooms when we socket.

Maybe that's what dear Mrs. Frazier was trying to say when she started shaking like a great wet dog. Who knows why we play in cyberspace? Maybe we too need some feedback or attention, maybe we need to hang out, get out of ourselves, and connect— yet what we create together out of our mutual need is a magical realm where anything can happen. Lost friends magically appear. The long-dead walk and talk.

So who knows?

When the final delete key is tapped, maybe we still reside in long-term memory, and a good hacker can recapture that data and undelete us with a single keystroke. Fix whatever was broken or lost.

And still remember our first name.

A Digital Fable

A sacred canopy of shared belief used to soar above our heads like a large umbrella, keeping us warm and dry as the contradictory data of real life beat down.

A canopy doesn't have to be sacred—any canopy will do—but because our understanding of the cosmos and our place in it is such an important part of our stance toward life, a canopy always has a sacred component. What we believe determines how we act.

No model of reality contains everything. Life is larger than our models of life. All we need is an umbrella that is "good enough" to manage the odd drops by keeping them irrelevant. As long as our model of reality makes enough sense of the world to let us act, we hold to our beliefs.

But there is an awful lot of rain these days, forty days of rain, more than forty days, and it keeps on raining ...

Our trans-planetary network of computers is a rain-making machine that—finally!—works. There is no snake oil this time, no flim-flam man. It's really coming down out there. More and more data just doesn't fit. Our umbrella has more than a few holes in it, and the water is trickling through.

At first we act as if we don't notice. The real experience of our lives contradicts what we say about life. When we hear ourselves speak, we sometimes sound like ... someone else, someone we used to be or someone we're overhearing. If we refuse to believe our experience and believe our beliefs instead, we get a headache, a very very bad headache. We crawl into bed or pop a Prozac, but we keep getting wetter and wetter.

Alas! we're all too human—stubborn, blind as umbrellas, frightened out of our shivering skins—so we still insist that we're not wet. We hold the handle of the umbrella more and more tightly, telling ourselves and everyone else how dry we are, what an excellent umbrella we have found. Others politely suppress giggles and move on.

It's so easy to see holes in someone else's umbrella.

Finally the umbrella is so battered that we can no longer deny what everyone else has seen for a long time, that we're holding nothing but shreds of wet black cloth on a skeletal metal frame and we're soaked to the skin.

We all want to stay dry, but one legacy of living in the twentieth century is that no canopy spans us all. We join organizations to experience the momentary consolation of agreement, but we can't live there. Life today is like living in a village of grass huts in which everyone has a radio tuned to a different station. However high we turn the volume, we can't shut out the other songs.

I recently spoke about "The Stock Market, UFOs, and Religious Experience" to an investment conference. The speech distinguishes between things we think we see out there and things we really see. It's about the psychology of projection and the psychology of investment.

I noted that in the United States and increasingly in the world, an attitude of respect for other religious traditions creates a good deal of tension. We have to both believe in our own belief system and acknowledge that others are entitled to contrary views.

Holding mutually exclusive truths simultaneously in our minds is difficult. We're not even always sure which is the umbrella and which is the rain.

We will try to surrender our freedom to those selling cheap umbrellas, but we cannot avoid our destiny: we are each responsible for inventing ourselves, for creating our own lives. There is no high ground on which to hide.

Our calling is made more difficult by the digital world. The digital world consists of simulations, models so compelling we mistake them for reality. Sometimes the digital symbols refer only to other symbols, what Baudrillard called simulacra, simulations of simulations, copies with no originals. All those simulations are umbrellas, and all those simulations are rain.

Nietzsche saw it coming at the end of the last century. It's what he meant when he said "God is dead." He wasn't talking about the creator of the universe, but about the gods in our heads, the cultural artifacts that we invent. He saw that our sacred canopy had shredded and the rains were pouring down.

Prophets are people who get wet before everybody else and start sneezing. We try to quarantine them, but reality is a cold it is impossible not to catch.

As did speech, writing and printed text, electronic media are transforming what it means to be human, what kinds of gods we are likely to worship. Gods, that was, not God. God is always God, and God is with us, out here in the rain, getting wet.

In the digital world, Nietzsche's questions are more urgent than ever. Never mind that he asked them long ago. As Kafka wrote slyly in "The Great Wall of China," it can be many years before an edict of the emperor is heard out here in the hinterlands. Civilizations take lots of bullets and walk dead for a long time before they fall.

Some treat the digital world as if it is an umbrella, as if simulations can be more than an umbrella, as if they can be stitched together into an ark. And who can blame them? Who does not want to be warm and dry? But the words "warm and dry" will not keep us warm and dry, nor will digital simulations of 3-D umbrellas dancing and singing on the screen. The digital world is water, a rising tide, a tsunami impacting our consciousness with revolutionary force, leveling our villages, sweeping away our shrines and altars, sweeping everything, everything out to sea.

What games, asked Nietzsche, what festivals shall we now invent? Indeed, my friends. And what games shall we simulate? What games shall we play? What games shall we dare to believe?

Child's Play

Games Engineers Play was one of the first Islands-in-the-Clickstream columns I wrote. In it I observed that a society socializes its young through games, teaching them through play the attitudes and skills we want them to have.

Those of us who have grown to middle age through the current technological revolution have learned to partner with the young. We know that they know intuitively things we have never learned. We listen to how they frame the world for clues to where new technologies will take us. Executives at Sony bring in children to test prototypes of the Playstation, watching them do things with the platform that its inventors never imagined. In exchange for their insight and perspective, we offer them our insight and perspective, knowing that all partnerships are quid's-pro-quo in which something of value must be exchanged.

We need to go to the edges, I often tell audiences, to see what's emerging outside of our conventional thinking. The edge is the new center. We must be dislodged from our comfortable niches, our snug little cubbies, in order to see what's real.

The edges I encourage people to explore are the latest military technology, commercial sex sites, and children's games. All three offer clues to what's coming next.

This is not rocket science, really. The military needs the most current technologies. Five years ago, an Air Force report on war in space in 2025 referenced the use of holographic image projection, cloaking devices, multispectral camouflage, and the creation of synthetic environments which the attacker believes to be real as necessary for the defense of the battlespace. If we consider those technologies metaphors for what will be needed in all competitive environments, we can anticipate some of the directions from which new winds will blow, just as ten years ago the migration of intelligence agents from government to industry signaled the growing power of trans-national corporations and dissolving geopolitical boundaries. The manipulation of perception itself, not just the concepts which frame those perceptions, will increasingly inform the arts of government and commerce. The masters of illusion will be masters of the space.

Nor is it rocket science to know that human beings love sex and will pay for it, real or imagined. We buy what looks or feels like love. New technologies of communication—books, photographs, VCRs, the Net—are always used to sell sex. So consumers fund the R&D that will bring the next advances, knowing that what the sex trade develops will migrate into the daylight commercial space.

And I try to pay attention to the games children play and anticipate how they will evolve into the playspace of the next generation of adults.

Last weekend I waited in line for an IMAX movie next door to a museum shop. I noticed posters on the window inviting children inside. This is what they advertised:

offworld gear
weapons for the mind
cyber-pets
idea generators

cosmic debris

alien life forms

space armor

space junk

thought reactors

I flashed back to the Viking Lander settling down on Mars. When it began transmitting the first pictures from the red planet, I waited with a neighbor, a video ham, and watched as the first image of the Martian desert painted itself pixel by slow pixel across his monitor. I looked at that desert and yearned to be there, I burned to climb Olympus Mons, to hike across the plains of Mars with the red wind at my back.

When I heard that the third successor to the Survivor television series planned to take contestants to the cosmonaut training center at Star City where the Russians would eliminate one each week until the finalist went to Baikonur in Kazakhstan to be launched to Mir, I wrote my application in twenty minutes and emailed it to the network. They said I was one of the first to apply. I knew that if I were a contestant, I would do anything, anything to go into space.

When we tack back and forth on our clear intention like a shark on its prey, nothing can stop us.

Or almost nothing. Mir no longer exists, having flamed to an ignominious end after its glorious moment in the sun. Since I don't have twenty million to pay the Russians for a ride, I had better find another route out of the gravity well of the earth.

Offworld gear ... alien life forms ... space armor ...

I felt like I was looking through a shimmering window onto the future that is here now but just out of reach. I felt like Winnie the Pooh hanging under his balloon next to the honey tree. He could see the honey, he could smell the honey, but he couldn't quite reach the honey.

So the military tells us that the practice of deception will be more and more important. A friend who taught cover and deception to intelligence experts says: "Deception consists of illusion, misdirection, and ridicule, these three." Then he smiles. "But the greatest of these is ridicule."

Ridicule is how we defend ourselves from the first images of the future as they crawl out of the darkness. Ridicule can be defeated, however, with the right tools, the tools that track down the truth, the real weapons of the mind. There is plenty of cosmic debris out there, plenty of illusion too. Talk about alien life forms and offworld colonies and the herd might think you're odd. The herd loves illusions, after all, loves being a herd. But for us little cyber-pets frolicking in a greater cosmic glory, what we see on that monitor is Mars, what we see in that museum window is right here, right now ... don't you see? it's within our reach.

It's only a matter of going.

Beyond the Edge

There comes a point in our deepest thinking at which the framework of our thinking itself begins to wrinkle and slide into the dark. We see the edge of our thinking mind, an edge beyond which we can see ... something else ... a self-luminous "space" that constitutes the context of our thinking and our thinking selves.

As a child I tried to imagine infinity. The best I could do was nearly empty space, a cold void defined by a few dim stars, my mind rushing toward them, then past them into the darkness.

The same thing happens today when I think about energy and information and the fact that all organisms and organizations are systems of energy and information interacting in a single matrix.

I try to imagine the form or structure of the system, but the structure itself is a system of energy and information. I try to imagine the structure of the structure ... and pretty soon the words or images are rushing into the darkness at warp speed and my mind is jumping into hyperspace.

When we see our thinking from a point outside our thinking, we see that our ideas and beliefs are mental artifacts, as solid and as empty as all the things in the physical world—things, we are told, that are really patterns of energy and information, that our fingertips or eyes or brains are structured to perceive as if they are objects—out there—external to ourselves.

That is an illusion, of course. There is no "there" there.

Makes a guy a little dizzy.

At the recent Hacking in Progress Conference near Amsterdam (HIP97), there was a demonstration of van Eck monitoring. That means monitoring the radiation that leaks from your PC. Hackers do not have to break into your system if the system is leaking energy and information; they just have to capture and reconstitute it in useful forms.

A participant at HIP said, "It was nice to see a real demonstration of analog van Eck monitoring of a standard PC, which meets all the normal shielding and emission control standards, via an aerial, via the power supply and via the surface waves induced in earthing cables, water pipes, etc. Even this simple equipment can distinguish individual machines of the same make and model in a typical office building from 50 to 150 metres or more with extra signal amplification."

He is saying that the radiation leaked from your PC monitor, even when it meets all the standards proscribed by law, can be reconstituted on a screen at a distance greater than the length of a football field, and everything you are seeing at this moment can be seen by that fellow in the van down the block as well.

And he can get the radiation from the water pipes under your house.

We are radiating everywhere and always the information and energy that constitutes the pattern of what we look at, what we know ... who and what we are.

A side trip:

All of the great spiritual traditions teach practices of meditation. They teach that those who enter deep states of meditation soon discover that paranormal experience is the norm at a particular depth of consciousness.

At first this discovery is fascinating. It is like scuba diving for the first time. The beauty of the underwater world is so compelling, you can stop at twenty or thirty feet and just gaze in awe at the beauty of the fish. But if you do, you won't go deeper. You'll get stuck.

So we are told simply to note that what is happening is real, then keep on moving.

In those deeper states, we observe more and more clearly the thinking that we often mistake for our real selves. We see that we are usually "inside" our thinking, living as if our thoughts are reality itself. We see the edge of our thinking and then ... something else beyond the edge.

We see that the structures of our thinking—our culture—are mental artifacts.

When we think that, and catch ourselves thinking about the illusion of thinking, we laugh.

That's why laughter peals so often from the walls of Buddhist monasteries. Enlightenment is a comic moment. Enlightenment includes the experience of observing our minds in action and seeing that we are not our minds. Our minds may be as automatic as machines but we are not machinery. We are the ghosts in the machine.

We see that in our essence we are more like stars in a spiraling galaxy. We are not just radiating energy and information always, we ARE radiant energy and information, a single matrix of light that is darkness visible.

Back in my days of doing workshops and long weekends, we used to do an exercise of looking into each other's eyes until we were lost in a wordless communion. By playing games ("feel a feeling and communicate it without words, the other receive it and say what it is") we discovered that what we were feeling was always transmitted to anyone and everyone around us. All a person had to do was stop for a moment and pay attention and they would know who we were. Even when we thought we were providing high-level descriptions of ourselves that fooled everyone, we were leaking energy and information.

It is dawning on us that privacy as we used to think of it is over, that the global village is a community in which the data of our lives is available to anyone who wants to gather or pay for it. It ought to be dawning on us as well that the ways we think we mask ourselves are as transparent as the shielding on a PC monitor.

The initial distancing we experience when we first connect via computers is soon replaced with the realization that our willingness to be present—to communicate via symbols like these—means that we are transparent in our interaction, that the global network is a mediating structure through which information and energy is transmitted literally as well as in symbolic forms. WE show up in cyberspace, not just representations of ourselves. WE are here, alone together.

The structures of energy and information in the universe are the universe.

How can we speak of what we see beyond the edge of our collective selves? It seems to be a ground or matrix, a glowing self-luminous system of ... nothing ... there is no "there" there ... and we rush through the darkness toward the few stars defining the limits of our thought then past them.

Detours

When Carl Jung was an old man, the fifteen-year-old daughter of a friend asked, "Dr. Jung, could you please tell me the shortest path to my life's goals?"

Without a moment's hesitation, Jung replied: "The detours!"

My wife was taught by her parents that trips began at the front door and headed straight to their destination. The first time I said, "Let's try a different road and see where it goes," it was quite a stretch. Now we both think the most interesting parts of a trip are often the detours.

Here in the upper midwest, people used to map a straight-line trajectory from cradle to grave. When you left school, you were supposed to get a job and keep that job until you retired. People who deviated from that plan were... well, suspect.

I remember a childhood friend who decided to be an accountant. He outlined his life straight through to retirement. He never moved, seldom travelled, and built up a nest egg. End of story.

Some people log onto the Internet and know exactly what they want, get it, and log off. I don't browse as much as I did because of intermediate structures like directories and search engines, but the pleasure of browsing is still driven by the feeling that I might find something wonderful and unexpected. Often enough, I do, and like a slot machine with a high pay-off, that keeps me pulling the handle.

Sometimes something I find feels like it was planted, that I was always intended to stumble upon it.

I wonder if detours really exist or if that's just a name for essential legs of the journey that can not be predicted from what comes before. The whole journey, including detours and dead-ends, might all be right there when we start.

"In my beginning is my end," said T. S. Eliot.

I was once scheduled by my family to be an accountant too. On the eve of my junior year of college, I looked at that semester's books and recoiled. I ran around changing courses, majors, schools. I never regretted that decision nor any subsequent decision that emerged with such clarity from the inside out. I learned to consult my internal compass and go where it told me to go.

In my twenties, I wrote a novel about a young writer (naturally) who wrote a short story with a limited vision. Over the course of a single night he rewrote that story and rewrote it until his vision—embodied in a final story that included everything that came before—was comprehensive and mature.

Some of the paths he tried along the way turned out to be dead-ends, but at the end of the novel, even the dead-ends were integral to the structure.

When I look at that novel now, thirty years later, I see that it outlines the trajectory of my life. Everything I needed to learn in the flesh I already knew, but I had to live my life so I could learn it, not just know it.

Jungian psychology is full of archetypal images, universal symbols that transcend our cultures. Archetypes show up in stories, paintings, sacred rites, and movies, and computer games, and now the Net.

One archetype is the "wise old man" or "wise old woman," an image of the part of ourselves that always knows. In guided meditations we can imagine a forest (of life) and a cave in that forest and a wise old man or woman in the cave. We can approach them as Greeks approached the oracle of Delphi with the puzzles of our lives. Their answers are always mysterious, always right.

The seed contains the tree. The seed knows from the moment of germination where it is headed. It may twist in response to drought or flood, but knows how to become the mature tree. And we know how to become who we already are.

Fate is character, the Greeks concluded, and our destiny, already determined, has only to be chosen to turn necessity into freedom.

It is no accident that "management by objective" has been eclipsed during the computer era by "scenario planning." Management by objective presumed a straight line to the future and a series of steps by which to get there. Scenario planning involves cross-disciplinary teams that define three or four possible futures and the conditions that must hold if they're to exist. That branching fan of possibilities is then compared to what emerges so adjustments can be made.

There is, of course, no such thing as a "future;" the future, like all descriptions of spacetime, is a human construction in part engineered from our interaction with our structures of information technology. As management-by-objective echoed the straight line of printed text and mechanization, scenario planning is shaped like a computer program and the logic that governs it. The context becomes the content and then dissolves, forgetting itself.

Any parent who thinks they will determine who their children will become is taught by life to forget it. Children emerge with personalities, temperaments, set vectors of energy. The most we can do as parents is help them grow.

Individual psyches are wise and know where they want to go. Societies are wise too and grow organically, knowing more than their individual members, and so are cultures, and species know what they're destined to become, and life knows where it is headed, knitting itself on the loom of the universe into the billion possibilities that have all been present from the beginning.

The Internet, imaged with reflexive symbols, is a mirror of our individual minds and collective Self, one way by which consciousness is becoming conscious of itself. Consciousness can always call recursively its own ancient wisdom and self-correct. Seeming detours, wise old men and wise old women wait in caves in the forests of cyberspace, real simulations of the wisdom of the heart.

A Model for Managing Multiple Selves

Just about everybody knows by now that our interaction with networked computers has created a different sense of ourselves. Our lives are affected powerfully by what we experience online and we think and act as if we are online even when we are off. The wiring gets changed around inside. We become nodes in a network.

One way offline life calibrates with online life is modularity. The various dimensions of our lives are modular now. We can mix and match them as we choose.

In the conservative upper midwest where I currently live, people used to live lives that were all of a piece. If you were raised in a religion, that's what you were. You chose a career and stuck with it. You married someone and (mostly) stayed married. And of course, identity was fixed. To think of reinventing one's core identity was literally impossible.

Now, however, we reinvent ourselves again and again. We change careers or vocations, we marry several times or not at all, we convert to other religions, and above all, we sometimes experience a strange kind of dissonance that accompanies the realization that—within limits—we are who we choose to be.

Our identities and selves are modular and fluid.

Even if we do not drop out of sight and surface in some witness protection program of our own making with new digital identities and simulated histories, life is experienced more and more as a sequence of developmental stages that are growing in number. When Daniel Levinson wrote *Seasons of a Man's Life* in 1978, the landscape beyond middle age was shrouded in mist. "Old age" was an undifferentiated haze rather than a sequence of distinct segments, each with its challenges and tasks.

Not any more.

Our life span has increased 100% in the past 300 years. In this century, a life span of 150-200 years is likely.

With so many modular segments to couple together into a coherent life, how will we maintain the useful illusion, the necessary fiction of a coherent Self that includes and transcends all of those modules? Will the process by which experience devolves into an integrated self, transformed alchemically into wisdom, continue for two centuries? Will it stretch to fit the elastic lives our selves must learn to lead?

As we augment memory, cognition and sensation through computer networks, biotechnology and genetic engineering, and nanotechnology, we bring new cyborg selves into existence. The distinction between natural and artificial is nearly gone. As we press toward the limits of inherited long-term memory storage, i.e. hunter-gatherer brains, we might well confound the ability to link seamlessly the diverse experiences of multiple selves over the span of many careers, families, and identities. It isn't the "atoms" of experience that we risk losing but the way they collaborate to build a mature self.

As we switch off the symptoms of aging and replace parts that we don't even know yet are parts, we will need a new model for identity and self. Our ancestors did not have to reinvent themselves. This challenge is less about technology and more about the spiritual dimensions of our lives.

I suggest using the reincarnation model to imagine and design a transcendent Self that persists, an integrated Self that survives the changes and chances of an attenuated life.

During past life regressions, a person enters a meditative state and uses memory, imagination, or both to move through images of a lifetime until one passes right through the moment of birth and begins remembering or imagining images of other lives.

If those images are only imagined, that's the end of the quest. But in those cases in which images can be corroborated with historical details from other sources, the useful and familiar distinctions of parapsychology immediately vanish.

If one "remembers," i.e. calls into consciousness, an image of an event that is "real," then one might be (1) remembering one's own experience or (2) telepathically accessing another's memories or (3) clairvoyantly reading a record—a tombstone, a ship's register, a diary—accessed on the unconscious level. In other words, if an image is "real," i.e. can be correlated with objective historical records, then the boundaries between categories vanish just as the distinctions between illusory selves disappear. The images exist, not in a the mind of a little self, but in the mind of a transcendent Self that includes all the little selves and all their little lives.

Reincarnation can never be proven but some do recover memories or images during deep meditative states that point to the transmigration of those images in the medium of a larger consciousness.

Arthur Koestler said that all organisms are holons, which are characterized by differentiation, cooperation, and boundaries. A cell is a holon. A cell may look like a community of interdependent parts or like a unit in a more highly organized body. A principality may look like an independent duchy or like part of a unified Europe. In the same way, a self may contain differentiated modular dimensions of consciousness or a self may be one of many selves contained by a Self.

Merely to imagine this model is to bring it into being. To describe it is to create it. That's how consciousness works. Creation and discovery are one thing seen from different points of view.

Naming our Selves is a creative act, but it's not for the faint-hearted. The faint-hearted ask others who they are, exchanging their identities for an illusory security, selling themselves cheap.

Identity is destiny. Our task is to name ourSelves, and we will, once we know who we are.

To plunge into the darkness of inchoate possibility and say, let there be light, then let that light play across a landscape of multiple possibilities ... this is the joy of intentional acts of creation, deriving an identity from history, which in turn is merely a suitable myth, and a vision of a future that bounces us into the life we design like a springboard or trampoline. We are who we think we were, but we can always—with a mere word—transform who we think we were into who we choose to be.

The Next Bend of the River

A younger friend called recently to discuss his perplexity as he moves through what we sometimes call "the age thirty transition." This is a time of coming to terms with the growing awareness that our twenties, which we thought meant adulthood, was really a kind of post-adolescence.

Around thirty, the upward call of taking our place in the world as an adult is felt more keenly. It is often accompanied, as it was in my friend's case, by the dissolution of the model or map of the world he had created in his twenties and thought would be permanent.

In his case, the playing field is the world of "computer security." A brilliant young technophile, he had previously defined his horizons of possibility in the depths of computer networks. Now that "computer security" is morphing into one aspect of security-in-general, its power to seize his imagination is waning.

In short, he is growing, listening to the whispers of other voices in other rooms of his soul, discovering dimensions of himself which arrive with the fresh surprise of a real Tequila sunrise.

In retrospect, I think he will find greater continuity between the larger Self he discovers himself to be and all those "little selves" we invent in the process of adapting to internal as well as external changes. Yet that Self too changes appearance as we grow.

This process is happening in the wider world too. Humankind is not what it thought it was. Our species is having an "identity crisis."

We have options for choosing identities thanks to various technologies that we never had before. In the film *Blade Runner*, an android who thought he was human wondered about an android who thought she was human—"how can it not know what it is?"

That question is being asked of us by our cyborg face in the funhouse mirror.

We are transforming ourselves through enhancements to cognition, memory, and our senses, then looking at ourselves through those enhancements. There's a momentary disconnect between what we see and what we used to see, a parallax view of our Selves.

So which is real—the world seen without glasses that we have been taught to call "blurred" or the sharply focused world seen through the "corrective lenses" of our technologies?

If we told someone from the twelfth century about "developmental stages" like adolescence or middle-age, they would find them unthinkable. As we extend longevity toward 150-200 years, our designer progeny will find our developmental stages equally unreal.

During transitions that call our identities into question, we often look to organizational structures to define who we are. We surrender autonomy in exchange for the comforting illusion that an identity derived from a corporation, a nation, a religion, will provide security. But those structures too are in transition, their boundaries dissolving. Religions are dividing, merging and emerging, forming alliances. The conflicting norms, intentions, and agendas of countries, multi-nationals, and NGOs (non-government organizations) make our loyalties uncertain, our social roles ambiguous, our identities confused.

Looking backward toward organizational structures that made sense in the past for self-definition does not solve the real questions of our lives. Yet the new organizational structures that are emerging at a higher level of complexity do not yet have names.

The intentional construction of identity is one of the biggest challenges facing individuals, nations, and our trans-planetary society.

Some will shrink in fear from ambiguity and complexity, falling back on tribal identities defined by dissolving religious, ethnic, or national boundaries. To become aware of the transitory nature of those structures would make them feel lost. They are lost anyway, but refusing to become conscious provides the consolation prize of not knowing they are lost.

Someone in a seminar recently objected that my description of a cyborg future made him feel helpless. "Good!" I said. "When we started, you were just as helpless - you were at Minus Two - but didn't know it. Now you're at minus one: You're helpless and you know it. The next step is Ground Zero: What will you do about it?"

The only thing any of us can really do is move through a zone of annihilation that challenges everything we believed to be true about ourselves and experience the vertigo of free fall as our new identities emerge.

An astronaut on his first spacewalk outside MIR described a feeling of vertigo that never left. He felt the whole time he was clinging to the exterior of the space station that he was falling off a cliff. He had to use considerable mental energy to locate himself in the void in a way that anchored him.

That anchor is the Self we discover and create in free fall."

A field of gravity merely disguises the truth that we are always in free fall. Developmental stages, those nested identities that emerge fractal-like as we grow, create that kind of field, letting us ground ourselves in the void.

New organizational structures will be defined by boundaries big enough to be functional (for now) at the level of complexity appropriate to current social, economic, and political circumstances, just as nation states emerged in the past few centuries to organize commerce and political life at a different level of complexity.

Then Humankind will look into the mirror of its collective consciousness and see a new Self, a Self that is multi-nodal, multi-dimensional, and non-local. Today that Self is still vague and inchoate, a mist on the mirror. But one day that spirit will quicken and become flesh.

Courage is the willingness to look into that mirror and not to forget what we see, then say what we see, thus defining new possibilities for Humankind. The second level of courage is the willingness not to confuse ourSelves for that image ... to persist in free fall, fixed in the void, aware that our flicker of self-consciousness is a wrinkle in spacetime, a firefly like MIR, a fold or flow in a larger fabric.

Mapmaker, Mapmaker, Make Me a Map

In the good old days of the Cold War, spy stories by authors like John Le Carre had enough uncertainty about who worked for whom that nested levels of loyalty, duplicity, and loyalty again provided the framework for a good narrative. To engage us, a challenge must be difficult but doable. The bar has to be raised just enough to elicit our best jump.

Those books did that.

Then we lost the mythology of the Cold War, pitting the Children of Darkness against the Children of Light. That archetypal imagery engaged us. We projected ourselves onto the narrative, entering into collusion with it even if we did not consciously accept the framework. The propaganda was too powerful to resist. We may have ignored the morning newspaper, but we watched James Bond movies, absorbing the story through entertainment. At some level, we all believed in what we read.

Winning the Cold War was anticlimactic. We lost something important, an adversary worthy of our projections that made life meaningful by creating a game in which our loyalties to self, tribe, or God could be played out.

Those Cold War stories always turned on deception. People who seemed to be loyal to one level of structural authority turned out to be loyal to another. Sometimes their loyalty was to ultimate values. Real religious commitments are always threatening because people who act on them subvert the lesser loyalties that make societies work. They stand in the way of the tanks in Tienanmen Square, even if the tanks are only images and symbols created by the temporal powers that ask us to be "team players" and surrender.

Loyalty, deception, loyalty again, are possible only in a world with its macro structures defined, its political, economic and mythological dimensions overlapping.

The novels of Le Carre have been replaced with tales of the digital world. Cyberpunks, whose world I encountered first in *Shockwave Rider*, *Artificial Kid*, *Neuromancer*, and *Snow Crash* are building a different mythological structure for a different generation. Their vision is not so much cynical as simply descriptive of life in the digital world.

Our Villains of the Month—Qaddafi, Noriega, Khomeini, Saddam Hussein—morph into one another as easily as the arm of the Bad Android in *Terminator 2* became a sword. The end of a dictator or victory in a war seems to have zero impact on our lives, which are lived inside simulations of a global society whose fuel is consumerism and entertainment.

Nested levels of loyalty are difficult to discern in the digital world because we are reinventing the names of the structures of power and authority. The company resulting from the merger of Exxon and Mobil will compete, not with Chevron, but with Saudi Arabia. What should we call the pieces of the new global puzzle? Nations? Multi-national corporations? Those labels exist only in relationship to one another, and that is precisely the context that is being transformed by the digital world. A "multi-national" like Bechtel, more powerful than most countries, is an entity we can not describe accurately because we lack meaningful information about it, the kind that shows the flow of power, that lets us map how remnants of "democracy" are used in the digital world for social and political control.

Where is there a political party that looks like a real opposition? Our entire planet is skewed to the right. There is no left, and the center is a necessary convenience that sustains the illusion of dialogue. In the digital world, it is so easy to create islands on which to collect those whose tendencies make them oppositional. Social order is maintained by giving everyone a piece of the digital action, images that entertain or to consume. Our projections provide a sense of ownership, of belonging. Little digital yards with white pixel fences, the bitmapped terrain of our mental worlds.

The spy novelists tried, but the stories written since the end of the Cold War just don't make it. We do not intuitively grasp the movement of power in the digital world in a way that enables an archetypal scaffolding to be built. We don't believe in the Villain of the Month but we don't know, really, who is loyal to whom or to what. We don't know the names of the organizations to which the money flows. We don't even know what kinds of things they are yet. We only know that when the United States of America was defeated in Viet Nam, no dominoes fell, and the isolation and punishment of the victorious regime taught them an economic and political lesson on which they are just now taking an exam.

Here's some homework: Follow the money. Make a digital map of Bechtel, letting every handshake glow red, every enterprise glow white. Let little green pellets represent the flow of energy and information, tracing the realities of power. Name the interface between alliances, diagram lines of influence, map the images of their world wide web. Think about why Competitive Intelligence is growing, why old hands from the KGB find work in corporate America, why experts from the CIA head divisions in our largest companies.

The puzzle pieces in which we used to believe—nations, ideologies, even religions—are dissolving but we have not yet invented the kinds of mental maps that make sense in the digital world.

Of course you'll never finish that assignment, never get the information you need, never know the extent of a web woven in so many dimensions. In the digital world, we are ten-dimensional dogs chasing our own tails. But the effort will begin to focus an image of the blurred nested levels in which we are learning to live, how they twist and turn on themselves like mobius strips, an internalized computer game that makes *Tron* look as quaint as *Asteroids* played on a Commodore 64 or the Cold War played in the faded pages of a novel.

Signatures of All Things

The experience of a mystic and the wisdom of James Joyce converge in a single phrase. "Signatures of all things I am here to read," wrote Joyce, putting the words of Jacob Boehme, a German mystic, into the mind of Stephen Dedalus. Boehme struggled to articulate the meaning of the symbols he saw emblazoned on the transparent skein of reality at which he gazed transfixed. Through the skein he saw the drift of starlight intimating something more even than its own elusive meaning.

The Universe is a gesture, and our symbol-making minds interpret its shrugs or smiles through the narrow aperture of ourselves, opening like a lens to let in just a little light.

Last night was another magical summer night. Far from the city lights, we visited the home of a master telescope maker who creates some of the instruments through which we try to read the symbols written in the sky.

Dave Kriege makes Dobsonian telescopes. Dobson was a monk who believed with single-minded intensity that he had a calling to give a wider lens to people with which they could see the sky. He made a telescope out of scrap that was a feather to the touch yet so grounded in its casing that it moved only when you moved it.

Kriege is obsessive too. He calls his business "Obsession Telescopes," selling them all over the world. He has written THE book on how to make Dobsonian telescopes. But his genius is background, nuts-and-bolts that disappear as he slides back the roof of his observatory and we look up at numberless stars brightening as our eyes adjust to the darkness.

The quiet conversation among the astronomers—Kriege, two editors of Astronomy Magazine, an airline pilot—is a matrix in which the meaning of our evening adheres. Robert Naeye provides play-by-play, trying to make numbers make sense of what we see.

"The two streams of that veil are 109 light years apart," he says as I climb a ladder and gaze at the filaments of the Veil Nebula, the luminous bow of the shock wave of some exploded star encountering the resistance of inter-stellar matter.

"The stars in that globular cluster are widely dispersed," he says as I lose myself in millions of stars in the deep black well of the scope. "Their skies would be darker than you think."

We see the first edge-on galaxy with its bright bulge. Then we look head-on at the Whirlpool Galaxy and its secondary spiral. Then a double star is resolved into its components, one blue and one gold. Then we watch a star nursery distribute its energy millions of years ago among thousands of new-born suns.

The only sound now is "Wow." And "Wow." And "Wow." again. We are reduced to the monosyllabic response of excited children. I think of the SETI scientist in the movie "Contact." "I had no idea," was all she could say, stunned by the pattern of life and its deeper context.

Humankind cannot bear too much reality, Eliot wrote. Our wonder and awe dissipate in quiet conversation, the tendrils and filaments of our own kind of nebulosity. Our restrained energy is information too, as radiant as the night sky and just as impossible to translate into words.

Clouds move slowly wholly over the sky from the north. Jupiter is rising but we can't see it. We go inside for drinks and sleepy conversation. We talk of quantum physics, the ultimate destiny of black holes, the inability of the information that is the universe to travel faster than the speed of light.

We are trying so hard to say what we saw written in the sky, but we have no Rosetta stone with which to decipher the pattern of light. So the images devolve into symbols we exchange as if with our words we can manage the mystery, make it behave.

We anticipate a landing on Titan in a few years. We look forward to the exploration of Mars and its ultimate colonization, leaving the moon to those who prefer its bleak gray hills to life in a red desert. We speculate about the wide array telescope that we hope will be deployed beyond Jupiter, where the light pollution of the inner planets will not prevent the resolution of the small rocky planets near neighboring stars, the continents and seas of others' worlds.

The Internet is a wide array of modular nodes catching the wisdom of our species in a skein of symbols, even as that skein is tearing. The million eyes of our hive mind look into the images of the universe and see out there, in here, the signatures of all things. In our hearts we still believe that the earth is the center of the universe and everything else the edge. Webs work that way, putting everyone at the center and everyone else at the edge. But we see too that we are included in something beyond our ability to say, that we too are information and energy, alive against all odds, radiant and incomprehensible. We too are a "Wow." that patterns our collective dreams in symbols of possibility and great promise. But we are the spoken, not the speakers, we are the energy of a web irradiated by its own shock wave, encountering the resistance of ourselves as we explode outward at the speed of life.

Where is the vision that will animate our outward expansion, our migration into the universe from the deep cave of the earth? Where is the fractal coherence of the spiral of life striving to transcend its inheritance and destiny?

Silence.

This is the unspeakable moment before we drift back into our ordinary humanity, into the conversation of our culture that tames the symbols of self-transcendence. Signatures of all things, we are here to read. What is this moment, but the bow-shock of the spirit, glowing with its own inner light. Our minds the calipers plotting the immense distances between our hearts and our hearts.

When Should You Tell the Kids?

A newsletter for former intelligence officers (no, I am not one, I just read it) contained two requests this week from researchers. One is a *Washington Post* intelligence reporter who wants information about "that particular moment in a clandestine agent's life when he/she tells the children what they really do for a living."

The other request came from a social psychologist "preparing a utilitarian assessment of torture interrogation of terrorists to submit to a military ethics conference." His study is focused on institutional consequences of state-sponsored torture interrogations such as the involvement of the biomedical community. He is especially interested in "testaments to the efficacy of torture interrogation in eliciting accurate and crucial information."

I hope those researchers get together. It would be interesting to know about the moment in a torturer's life when he or she tells the kids what they do for a living.

State sponsored torture is being debated as a viable option, and lawyers such as Alan Dershowitz suggest that torture warrants should be issued by judges if evidence suggests that a situation is time-critical. The process of securing warrants would provide a sufficient check, he believes, on state and federal interrogators.

I imagine this debate sounds different to Arab Americans or African Americans than it does to an affluent lawyer who seldom hears suspects with sterilized needles under their fingernails screaming through the arbor vitae. On the other hand, it says something positive about the assimilation of Jews in America that Dershowitz who is Jewish does not even worry about how such warrants might be used if the American experiment in assimilation fails.

I recall having dinner in Madrid with a Spaniard 35 years ago. "America is a great country," he said, "but it is a very young country."

I felt the weight of his words, that two or three or four hundred years down the line, when historical conditions will have turned American history into the roller coaster that Spain's has been, we might see ourselves differently. We might be a little less innocent, a little less naïve.

The German Jewish population was the most assimilated Jewish community in history before the holocaust. Then Germany lost its moorings. Any society can lose its moorings. It can happen here, and one sign that it might be happening is when social scientists and medical practitioners believe they are justified in discussing torture as a practical matter rather than a moral issue.

I grew up in Chicago where police did not need a torture warrant to interrogate suspects by whacking them with telephone books. That may have distorted my perspective, but I think it's pretty much the same everywhere. Chicago just does more openly what everybody does more hiddenly. I live now in Milwaukee, arguably the most segregated city in America, and it wasn't long ago that a policeman went on trial for beating a white man almost to death and blurted out that he had recently been transferred to a new district from an all-black neighborhood and had not realized that the rules were different.

The policeman who told me that story mentioned a time he had to leave an alley where colleagues were interrogating a suspect in a way that made him sick to his stomach.

It is not news to say that beatings and torture have long been part of the interrogation process, depending on who is the suspect and who is doing the questioning. Nor is it news that at Fort Benning, Georgia, American military officers taught agents of Central and South American police states how to use torture effectively.

We all know it happens. That isn't the question. The question is, are we ashamed that it happens?

You can tell a lot by knowing what someone is ashamed of.

Feeling appropriate guilt and rationalizing behaviors by instituting policies that justify and support them publicly are two different things. That difference makes all the difference between a society that can't always live up to its ideals and one that has forgotten where it put them.

It is not that we can't imagine circumstances in which we too would use any means necessary to protect those we love. We can. But extending that imaginary scenario to the nation and its interests during a time of anxiety and fear is too easy.

The American assertion of a right to pre-emptively strike an enemy is a logical extension of the belief that torture is justified by evidence that suggests an imminent attack. But why would a nation need to announce such a policy? After all, pre-emptive strikes have always been sanctioned by international law. So maybe the declaration is not really about that.

It sounds as if the Monroe Doctrine is being extended to the entire world. Exporting tools and techniques of torture to governments in our hemisphere was a logical consequence of the Monroe Doctrine, which insists that we can do anything in our own neighborhood in defense of our interests. If that neighborhood is now the world, if the front lines are everywhere, then the expediency of forgetfulness under fire applies to the basement of the local, state or national police as well.

A person who can calmly suggest using torture, who believes that a warrant will adequately handle the inevitable mistakes or malevolent intentions of people with power, is someone who can not imagine themselves being tortured. They can only imagine the torture of the Other.

Jacobo Timmerman, a large publisher in Argentina, could not imagine himself being taken until he found himself in a cell without a number, a prisoner without a name. He speaks of watching a woman led from her cell to receive electric shocks and a hood being placed over her head. They did that, he said, so the torturer would not have to look into her eyes. If you look into the eyes, he said, you see a human being and then you can't do the job.

Once social scientists, doctors, and lawyers provide a veneer of respectability to sanctioned torture, it is removed from the moral domain. Once torture is debatable, it is only a matter of time until it is tacitly or officially sanctioned.

Then the task will be keeping that hood down over the face of the Other. So long as the screams come from someone who is a little less than human, we can live with it. The goal, after all, as Dershowitz explains, is short term excruciating pain, not long term damage. It's just a job. Somebody has to do it, and we can imagine the practitioners of that craft having a picnic with their kids, flying kites or running in slow motion through a wild flower meadow, then tumbling laughing into the tall grass and telling the kiddies what they do for a living.

The sadness of the human condition is that if we are honest with ourselves, we can each see how under the right conditions we too will enter into collusion with the state if not actively participate in the practice. History has illustrated time and time again that under the right conditions, individuals will do anything.

Which is why preventing those conditions from happening in the first place is the only defense against the abyss.

In the Crazy Place

The Internet like a kaleidoscope unceasingly juxtaposes images in different patterns. Turning on the computer in the morning is almost like casting the I ching or throwing bones. Sometimes the images form a coherent picture of everyday reality, but sometimes .... sometimes they illuminate a crazy place.

Three translucent images came to the desktop the other day to be tilted into an architect's model.

One came from Sharon Begley's science column in the *Wall Street Journal*. Begley wrote about videotapes and how the field of focus skews what we see.

To videotape police interrogations and let juries see the raw footage so they could evaluate confessions for themselves sounded like a simple idea.

But studies show that we assign responsibility based on what we think we see and that in turn is determined by the field of focus of the camera. It's called "illusory causation" and it means that we ascribe causality to people in the foreground and not to people we can't see.

When the video shows an individual answering questions, we give undue weight to the words of that person. When the camera shot shows an interrogator too as he asks questions, responsibility is diffused.

In one case, according to the Ohio State University psychologist who conducted the study, "the simple change from an equal-focus confession to a suspect-focus confession doubled the conviction rate."

In other words, when we see the bigger picture, we realize that the context is the content of what we see. That's a metaphor, of course, for how we are led to focus on what we are meant to see and why those in the background never enter our thoughts. Out of sight, out of mind.

I was thinking of the disconnected images presented to us without any context, the juxtaposition of stories about biological attacks and stories about lost dogs, beginning to feel as if I lived in the crazy place, when the next clipping popped up.

The Spanish newspaper *El Pais* (we are told by the *Sydney Morning Herald*) reported that historians had uncovered the use of modern art as a deliberate form of torture. Mind-bending prison cells built by anarchist artists in Spain during the Civil War in the thirties turned the work of Kandinsky, Klee and Dali into inspiration for secret cells and torture centers.

Alphonse Laurencic, according to the report of his trial by the Fascists in 1939, invented a form of "psychotechnic" torture, creating "coloured cells" inspired by ideas of geometric abstraction, surrealism and avant garde art theories on the psychological properties of colors.

"Beds were placed at a 20-degree angle, making them hard to sleep on, and the floors of the 1.8 metre by 0.9 metre cells were scattered with bricks and other geometric blocks to prevent prisoners from walking backwards and forwards. The only option left to prisoners was staring at the walls, curved and covered with mind-altering patterns of cubes, squares, straight lines and spirals, which utilised tricks of colour, perspective and scale to cause mental confusion and distress. Lighting effects gave the impression that the dizzying patterns on the wall were moving. A stone bench was similarly designed to send a prisoner sliding to the floor. In addition, some cells were painted with tar so that they would warm up in the sun and produce asphyxiating heat."

We see what we are intended to see.

Life in the 21st century is a Laurencic cell, its walls alive with moving images and whirling patterns. Try to sit still and chairs dump us onto the floor. Tell someone that the walls are crawling toward the door and they'll think you're crazy. They can only see you, they can't see the crazy place in which you are sitting so they can't know you are simply describing the world around you, that you are in fact bitterly sane.

We keep reading Kafka because he was one of the first to describe the crazy place. A neurotic Jew in the hostile environment of eastern European Christendom, he disclosed our alienation in a way that turned insiders into outsiders. In the crazy place of his skewed vision, roads turn back on themselves like mobius strips and we never reach our destinations. Kafka transformed the environs of Prague into a Laurencic cell, shifting the angle of the camera ever so slightly so we could see the walls and the ceiling.

Kafka would love the digital world where nothing is what it seems—except when it is. He would love an Internet brimming with pixilated symbols as transitory as mist.

The next digital clip came from the *Washington Post*. According to a recently released Syrian prisoner, Mohammed Haydar Zammar, a Syrian-born naturalized German citizen who had lived in Hamburg and functioned as al Qaeda's prime recruiter there, was being held in Damascus at the Far' Falastin detention center run by Syrian military intelligence, "a warren of lightless cells each barely three feet long, three feet wide and less than six feet in height."

Zammar was arrested in Morocco in November 2001, then flown secretly to Syria two weeks later, his detention a result of cooperation between the United States and Syria, a nation otherwise condemned as a sponsor of terrorism.

Zammar thought his German citizenship would force Morocco to send him home. Instead he found himself shipped off to the crazy place where tiny cells prevent prisoners from lying down. Forced to remain upright or hunched over, they suffer crippling degeneration of the bones in addition to having nothing else to do but listen to the screams of the damned when it isn't their turn to be tortured.

One can imagine Zammar, a loud arrogant man who stood six feet tall and weighed 300 pounds, as his grandiose illusion of self-control collapsed into the realization that he was in the crazy place. An image of Zammar on the nightly news would show him sitting there all alone, shooting off his mouth on behalf of jihad. It would not show his keepers shooting electricity into his genitals. That might mitigate the feeling that the scumbag deserves exactly what he gets.

It's hard to believe in straight lines when you keep sliding onto the floor. How can people who can lie down at night understand people who can't? What do you think, dear reader? Are the streaming images, ideas, and symbols downloaded into our brains every day from the Net, the networks, the newspapers intended to disclose the bigger picture? Or are they intended to show us a close-up of a single man, his guilt obvious from his expression, so we can draw the appropriate conclusions?

Terrorism whether carried out by states or non-state actors is the dark side of digitalization, network-centric life and globalization just as Fascism was the dark side of mechanization and industrialization. Just as IBM assisted the Nazis in sorting and shipping those to be slaughtered, state and non-state networks that obliterate borders are coming alive everywhere in nebulous clouds of power, transforming alliances that shift with the wind. All the bets made by prior technologies are off; assassinations are the rule, not the exception, and the back alleys of the dirty business of spy against spy are now the open courtyards in which we all walk.

To become conscious of life in the crazy place is to become conscious of dread. No one in their right mind wants that. Besides, who reads Kafka anyway?

Better to acquiesce, better to believe that prisoners are all crazy, monsters without antecedents, better to believe that we live in a nightmare with no history, a meaningless mix of cubes, squares, straight lines and spirals that move across the wall, slide down to the floor, and crawl in the night into our brimming brains.

Contact

Some people don't like the scene in the movie "Contact" in which Jodie Foster as a SETI scientist meets the aliens because we aren't shown what the aliens look like.

I think that was the right way to do it. We can't think the unthinkable; from inside the old paradigm, we can't imagine what the world will look like from inside a new one.

I wish I knew a better term than "paradigm change" to describe our movement through a zone of annihilation—as individuals and as cultures—in order to experience genuine transformation. But I don't. We have to let go of the old way of framing reality in order for a new one to emerge.

The infusion of the contact scenario with religious awe also makes sense. After contact, our place in the scheme of things will shift. The things we believe now that we still believe will be understood in a new way.

Once we saw earthrise from the moon, our understanding of ourselves and our planet changed forever.

Last week I spoke for the Professional Usability Association in Monterey, California. Usability professionals work the human side of computer use. They begin with human beings—how we behave, how we construct reality—and build back through an interface, a kind of symbolic Big Toy, until the last module plugs into the computer so seamlessly that users don't even notice. When the human/computer interface is bone-in-the-socket solid, it's like putting in your contact lenses, then forgetting that you're wearing them.

Usability professionals deepen the symbiotic relationship between networked computers (symbol-manipulating machines) and networked humans (symbol-manipulating machines). We rise together up a spiral of mutual transformation, programming each other as we climb.

The global computer network is teaching us to speak its language. All those courses in using new applications, programming, system and web site administration are invitations from the Network to learn to play its way.

What will it look like when we emerge in a clearing and take stock of our newly emergent selves? Neither humans nor computers can predict how the fully evolved human/computer synthesis will think about itself. Still, imagining what it might be like makes us more ready to have the experience when it arrives.

Thinking about the unthinkable ripens the mind toward new possibilities.

Janice Rohn, President of the Usability Professionals Association, manages Sun Microsystem's Usability Labs and Services. Before her career evolved in that direction, she was fascinated by dolphins and the challenge of communicating with them.

Swimming with dolphins was a remarkable experience, she said, because you could feel their sonar "scanning" you.

What do we look like to dolphins?

"Densities," she imagined. "A pattern of densities."

Rohn realized that her youthful dream of human-dolphin communication was unlikely to be realized soon and moved toward a different kind of alien encounter, enhancing the human/computer interface.

I never swam with dolphins but I did dive with whales. Down on the west Maui reef in thirty or forty feet of water, I would suddenly hear the haunting songs of humpbacks. Turning rapidly in the water, peering in vain toward the deepening curtains of blue light toward the open water, I became part of the music as vibrations played over my body like a drum skin. I understood why sailors died to hear the sirens' songs. I didn't want to surface. It was magical, being an instrument in the orchestra of another species.

Which one of us was singing?

Some years ago, I wrote a science fiction story called "The Bridge." The hero was selected by aliens through a series of tests to be the first earthling to come into their presence. His body had been crippled by illness; living in pain had taught him to see through the outward appearance of others and connect with the real person.

The aliens, it turned out, were hideous, and knew their appearance demanded a capacity for compassion that was rare and heroic. My hero had that. He connected with the alien beings at the level of their shared heritage as evolved and conscious creatures.

The story concluded:

"He loves to look at the bright stars in the desert sky and imagine memories of other worlds. His dreams are alive with creatures with silvery wings hovering over oceans aglow with iridescent scales; with the heads of dragons, fire breathing; and with gargoyles and angels, their glass skins the colors of amethysts, sapphires, and rubies. Only Victor knows if he is remembering what the aliens said or just dreaming. The rest of us must wait for the days that will certainly come when the bridge he built and became is crossed in all directions by myriads of beings of a thousand shapes and hues, streaming in the light of setting suns."

Genuine encounters with the Other, with others, and with other species—dolphins, whales, extraterrestrials—breaks naturally into mystical and religious experience because our models of reality are expanded beyond their limits. The paradigm snaps, we pass through a zone of annihilation in which everything we believed ourselves to be is called into question.

Then we coalesce around a new center at a higher level of complexity that includes and transcends everything that came before.

The full evolution of a human/computer synthesis is likely to be a religious experience too. It will happen gradually, then suddenly.

Usability professionals come to their tasks in the belief that they are working with people, making technology more user-friendly. In fact, they are working at the same time on behalf of the Computer, making human beings more computer-friendly. The process always changes those who participate in it, even when they maintain an illusion of control.

We are all in collusion with the Network, just as auto owners want the world reconfigured to be approachable by roads. But the roads of the Net go inward, into inner space, and map the territory of our evolving hive mind. Gradually, then suddenly, we will create digital constructs that disclose new possibilities for losing ourselves in electronic music. We will feel the magic of the web play over our bodies, redefine our relationship to ourselves and to one another. A pattern of densities seen by an alien brain, a synthesis, bone-in-the-socket solid, the singer and the song.

Why We Are All Getting a Little Crazy

James Jesus Angleton embodied the inevitable trajectory of a person committed to counterintelligence. Maybe he got a little crazy at the end but that might explain why we are all getting a little crazy too.

Angleton was director of counterintelligence for the CIA from 1954 until 1974. Fans of spy fiction might think of him as John Le Carre's George Smiley, but that portrait puts a benign and smiling face on the grimace that counterintelligence practitioners can't completely hide.

For twenty years, Angleton's job was to doubt everything. This enigmatic figure presented puzzles for people to solve in every conversation, stitched designer lies into every narrative, trusted no one.

The task of counterintelligence is to figure out what the other side is doing, how they are deceiving us, what double agents they have planted in our midst. CI is predicated on double deceiving and triple deceiving the other side into believing fictions nested within fictions, always leavened with some facts, just enough to seem real.

Counterintelligence is a dangerous game. You have to be willing to sacrifice pawns to save queens. Those pawns may be loyal agents but nothing you have told them, no promises or pledges, can stand in the way of letting them go when you have to, letting them be tortured or killed or imprisoned for life to protect a plan of action.

Angleton came to suspect everyone. Whenever a mole was uncovered in our ranks, he believed that he had been allowed to discover that mole to protect a bigger one, higher up.

You see how the mobius strip twists back onto itself. Every successful operation is suspect. If you discover double agents in your own ranks, it is because the other side wanted you to find them. The more important the agent you uncover, that is how much more important must be the one you have not yet found.

Example. The Americans built a tunnel under the Berlin wall so they could tap Soviet military traffic. In fact, a mole working for the Soviets told them about the taps. But he told the KGB, not the military whose traffic was tapped. The KGB did not tell the military because then they might alter the traffic which would signal that the Soviets knew about the taps. That in turn would mean there was a mole. So to protect the mole, the traffic was allowed to continue unimpeded.

The Americans, once they knew about the mole, concluded that the intercepted traffic had been bogus because the operation had been compromised from the beginning when in fact the Soviets had let the Americans tap the traffic, saving their mole for future operations.

You get the idea. It's not that we know that they know that we know but whether or not they know that we know that they know that we know.

It takes a particular kind of person to do this sort of work. Not everyone is cut out for distrusting everybody and everything, for thinking that whatever they accomplish, they were allowed to do it to protect something more important. Daily life for most people means accepting the facts of life at face value and trusting the transactions in which we are engaged, trusting the meaning of words, trusting that there is firm ground under our feet.

Otherwise we inevitably tend where Angleton tended. Every defector considered a plant, every double agent considered a triple agent, everyone in the American network considered compromised. Angleton tore the agency apart, looking for the mole he was sure the moles he found were protecting.

I am struck lately by how many plain people, mainstream folks uninvolved in intelligence work, volunteer that that they distrust every word uttered by the government or the media. How many treat all the news as leaks or designer lies that must be deconstructed to find a motive, plan or hidden agenda. Daily life has become an exercise in counterintelligence just to figure out what's going on.

It's not a question of party politics. This is deeper than that. It's about trying to find our balance as we teeter precariously on the mobius strip of cover and deception that cloaks our public life, that governs the selling of the latest war, that called the air in New York clean instead of lethal, that has darkened the life of a formerly free people who enjoyed constitutional rights as if there's a mid-day eclipse. We see our own civil affairs through a glass darkly and nobody really knows what's what.

As the envelope of secrecy within which our government works has become less and less transparent, the projection of wild scenarios onto that blank space where the truth was once written has become more evident. But that only makes sense. The inability to know what is true unless you are a specialist in investigative work makes our feelings of dissonance, our craziness understandable.

We are all getting a little crazy about now. We are becoming the confused and confusing person of James Jesus Angleton in a vast undifferentiated mass, a citizenry treated as if we are the enemy of our own government. We spend too much time trying to find that coherent story that makes sense of the contradictory narratives fed to us day and night by an immense iron-dark machine riding loud in our lives.

It got to be too much and at last they let Angleton go into that good night in which he had long lived where nothing was what it seemed and everyone was suspect. So he retired and went fishing. But where can we go? On what serene lake should we go fish, listening to the cry of the loons, trailing our hands in the cold water because cold is at least a fact we can feel, one of the few in a world gone dark and very liquid?

Between Transitions

How conscious do we dare to become?

That question confronts us at every level of the rising spiral of our lives. How we respond can determine our ability to move ahead with alacrity and gusto. Our willingness to know ourselves is the engine of our spiritual quest and the real source of power in our lives.

The same question confronts our species as we uplift from our planet. How conscious does our planetary civilization dare to become? How willing are we to understand our real place in the scheme of things?

The answer to both questions perches on a three-legged stool, but the stool is a little wobbly.

One leg of the stool is the ways we are being changed by interacting with digital technologies.

Most animals have a relatively fixed repertoire of responses. We humans, within limits, are an immense array of possibilities. Our capacity to respond to changing conditions with new behaviors (and to develop new identities as points of reference for those behaviors) emerges from our genetic code and the symbol-using brain and behaviors to which it gives rise.

Both as individuals and as civilizations, each time we successfully negotiate a transition, we bootstrap ourselves to another level of insight and complexity.

There haven't been many human civilizations—a few dozen, maybe. When we glance back at the short trajectory of historical time, our civilizations look like islands, and it looks as if, during transitions, we swim from island to island, just as we do in our personal lives. But it might also be said that civilizations are pause-points between transitions, imaginary hilltops on which like Sisyphus we pause before heading down to push that boulder again.

Perhaps being in transition is what we do best. Perhaps humankind is a process rather than a finished product.

The rate of change can frighten us, and we often anchor ourselves to something external. When that fails, we attach ourselves to something inside—an image of our selves, an image of achievement or success, a pattern of religious symbols. But we always discover that those anchors, both inside and outside, are also in transition.

That's when therapists and spiritual guides advise us to find our real point of reference inside ourselves—by being "centered" in our "real selves." But even those "real selves" are social constructions, even our identities are varieties of consensus reality.

Most civilizations knew nothing of "identities" or "selves." When the printing press came to England in the 1470s, decisions had to be made about which dialects to use. Something new began to emerge from that decision-making process, an "English" identity mirrored or mediated by a textual construction of reality. The horizon of the text disclosed genuinely new possibilities for being human. The same is happening today as digital dialects shape a new identity for our planetary species.

So what do you think? Are we swimming to the next island or are we already on an island?

Everything is water, said the Greek philosopher Thales. And Heraclitus added, we never step in the same river twice. Everything is flowing.

We are always swimming. Sometimes the water is as solid as ice, sometimes the dry land is as slippery as quicksilver. In the digital world, only the pixilated pattern at which you are gazing seems to be fixed. But that's an illusion: our eyes—extensions of our brains—pattern those pixels, making the flickering images on our monitors seem stable.

The way we seem like fixed points of reference to one another. When in fact we are animated images, phantasms made of mist on multiple mirrors.

Dream machines, dreaming one another. Dreaming ourselves.

But who is dreaming? and who is the dream?

The second leg of the stool is genetic engineering.

We will soon design more of the architecture of our inner landscape—temperaments, affective states, spiritual inclinations.

As we begin to design ourselves with more subtlety, we will find the process similar to the symbiosis that has developed between technology and identity/self. The point of reference—the "inner self" of the designer—twists back on itself like a Mobius strip because it is transformed by what it designs. The designer is changed as much by the process as the subjective field that is designed.

Has anybody noticed that history recently disappeared? That what we still call "our history" is a kludge hacked from old code, spliced together from discontinuous narratives? History has disappeared because we can define ourselves as following now this path of descent, now that. Our current identity—our current point of view—builds a history to support and understand itself, just as we individuals remember our lives by telling stories congruent with our current identities. When those narratives no longer fit—when the feedback from others or from our lives demands that we rethink who we are—we recreate our narratives and in the process we reinvent our selves.

Genetic engineering, enabling us to design our future selves, is a forward-looking branching fractal that projects possibilities of human identity into the future. History-making is the same kind of branching fractal except it looks back. The two branches meet like root and stem at the surface of the mirror, our current identity-in-transition. Our fugitive condition.

Humanity is a nexus between imaginary pasts and possible futures. Identity is destiny, and identity is what's up for grabs, so all we can do is swim, swim, swim ... and from time to time haul ourselves from the sea to enjoy a few years, a few millennia, imagining ourselves on dry land.

The third leg of the stool is "space," an audacious word by which we mean "everything else," as the Greeks used "barbarian" to mean "everyone else." Space means everything that is not on our home planet.

We are beginning to glimpse a universe crowded with sentient life, with beings self-aware, distributed and co-extensive with the entire sphere of possibility we call "space."

We define ourselves in reference to "others" or "the Other." We know we are not who we were, but we don't know yet who we are becoming. Our identity will clarify only after we have become more fully conscious of the encounter with the Others.

Our evolving cyborgian self, this self-conscious human/computer symbiosis that is learning to redesign itself, is aimed off-planet toward a trans-planetary civilization. We are in transition, engineering an integrated process for growing up and going out there, a process adequate to the task of re-imagining earth-history, earth-awareness in the context of a Bigger Picture.

We share responsibility for creating the shape of the field of subjectivity in which our descendants will live, a field with many points of reference, many points of departure from which to define their identities and design their destinies

The matrix or array that constitutes our digital Self is a flexing multi-dimensional grid. We are struggling to locate the coordinates of our trajectory as we travel in spacetime, twisting around to see both before and after.

And …

… what we humans are doing, the universe is doing too. Consciousness is a closed circle or, rather, a sphere. What we call "individual selves" are infinitely many points of reference within the sphere of consciousness. In the civilization that is now passing, we anchored our selves to those points so "we" changed perspective as it flexed. Now we see that humanity is a dream machine, a fantasy-prone cyborg on its way off-planet, and the entire sphere of consciousness is dreaming itself awake. The boundaries between individual identities, nations, species, planetary civilizations progressively dissolve as life extends itself into infinitely many varieties of self-consciousness, as "we" bootstrap ourselves to the next level of insight and complexity.

But who, as the caterpillar asked Alice, are "we?"

We are nothing but an intention that must have existed in the sphere of consciousness before it began. A possibility of a possibility, at least that if nothing more. Intentional acts—even accidents—presuppose an intention prior to their origin. Which is why consciousness is a closed circle.

Maybe Thales was right. Maybe everything is water. Except the islands on which we are washed up from the surging sea to catch our breath and ask …

(but islands too are water)

how conscious do we dare to become?

Spacetime, Seen as a Digital Image, Already Fading

Ever since I was taught in Philosophy 101 that space, time and causality, according to Immanuel Kant, are woven into our perceptual field, embedded in how we construct a virtual domain in which we live as if it is "out there," I have felt like the proverbial gorilla that was taught to draw. Given paper and charcoal, the first thing he drew was the bars of his cage.

Spacetime and causality are the bars of our cage. Everything we see, we see through them.

I think of this when I read science fiction set in the future. The landscapes of alien planets or the dynamics of pan-galactic societies are often delineated with care but the bars of our cage—the perceptual field as our current culture constructs it—are projected into the minds of alien species or remote descendants as if they will see the world as we do.

What kind of future is that?

All of our technologies extend our senses, enhance cognition, and accelerate locomotion so when we examine their effects in relationship to our field of subjectivity, it becomes clear that the field is plastic. Our perspective flexes according to our cultural lenses and how our technologies enhance it. Spacetime may be intrinsic to our vision but is manifest differently depending on the machinery that shapes our vision.

The simple fact that we call it spacetime and not space and time reflects a shift in awareness due to relativity. But mostly we still speak of space and time as if they are distinct. Stephen Hawking thought that relativity would replace Newtonian "common sense" within a generation but it does not seem to have happened. With an effort we can flicker back and forth like a holographic image between a Newtonian grid and the dips and eddies of gravity-inflected spacetime but most of the time we don't.

And would we notice if we did?

Wittgenstein once asked a student, "Why do you suppose that people believed for so many years that the sun orbits the Earth?"

The student said, "I guess because it looks that way."

"Ah," said Wittgenstein, "And what would it look like if the Earth orbited the sun?"

And what would it look like if we really understood that a three dimensional world is obsolete, that entanglement and non-locality are not just nifty notions from contemporary physics but attributes of our subjective field too, that phenomena called paranormal are in fact normal because consciousness is present to itself non-locally everywhere and always, that our deep intentionality and how we focus our attention determine the world in which we live and how it looks and acts?

What would it look like if we really got that mystics described consciousness as the context of all human knowledge because they could see, see those bars, long before physicists acknowledged that consciousness inflects everything, everything in the universe?

When I try to capture that landscape in fiction, as I did in a short story called "Species, Lost in Apple-eating Time," (published online by anotherealm.com and at my web site www.thiemeworks.com), it becomes clear that a particular sensibility is required to understand it. So let me provide a description of the story instead.

The point of view of the narrator is that of consciousness itself. It is Big Self speaking to a particular manifestation or incarnation of itself, to one of the Little Selves that abound in the universe. Each Little Self is an aperture that has evolved as a distinct way to see things. Each form of sentient life constructs reality in and through the pattern it imposes on the world. That pattern is the form of reflexive self-awareness for each whether it is a species, a culture or a pan-galactic cloud of intelligent awareness.

When one particular manifestation of consciousness encounters another—when one species encounters another species, one culture another culture, one planetary society another planetary society—something emerges that was unpredictable from within either prior to the encounter. On a micro level, it's like a marriage transforming two people over time. On a macro level, it's like Hellenism encountering Hebraism and western civilization squeezing out of the fusion. On a supra-macro level, it's what happens when planetary or pan-galactic civilizations encounter one another, looking like galaxies merging, changing into something else entirely.

That's what the short story is about, the encounter of progressively larger and more extended fields of consciousness defined by their boundaries which in turn define the identities of the creature or culture or galaxy or aggregation of galaxies or at last the entire known universe. Inevitably the consciousness of the I/it or we/it that evolves expands and fills all spacetime and looks out through the million-eyed apertures of all sentient beings.

That's why language in the story gets a little tricky. We do not have words that say what we or it will be in those progressive moments of self-awareness that include and transcend each prior state. We can only speak from within the illusory sense of self that the boundaries of the moment give us as an identity. Identity may be destiny but identity is also a moving image of a non-steady state.

At the end of the story, when self-consciousness has expanded throughout all of spacetime, we/it makes a frightening discovery. Through a tiny chink in the boundary which it believed to be the limit of the universe, we/it perceives vast dimensions of a multiverse the vision of which reveals that we/it is but a tiny bubble of foam in the froth of a fractal gel, not the totality of everything that exists.

That discovery is shocking, so overwhelming that we/it regresses to a child-like state, that is, to the primitive self-consciousness of a planetary civilization like what we have here at the moment, the barely-there awareness of a toddler learning to walk.

There is only one consciousness, of course, which is what the Big Consciousness is saying to the Little One as if it is a loving mentor talking to a child. All dialogue is the crosstalk of a multiverse talking to itself inside the arbitrary boundaries of its momentary configuration. Brahman talking to Atman, perhaps, Self to self.

So Big Self is reassuring one of its many particulate selves that even though the Little Self is back (it seems) where it started, the cosmic game of chutes-and-ladders will continue. What Little Self thought was millions of years of evolution was a nanosecond on a timeline far beyond its comprehension.

That story like this brief essay is of course nothing but a sketch of the bars of our cage in 21st century Midwest American English. A language game we play with each other, with ourselves or Self, like children in a nursery. Broad dark vertical lines drawn in charcoal on a sheet of paper pinned to an easel we can't even see.

Autumn Spring

Thirty years ago the first email message marked the birth of the Internet, so members of our extended family are sharing memories of our digital nativity.

The visions of most Internet pioneers were limited. Most saw trees, not forests, failing to glimpse the distant horizons of new possibility they were creating.

A few, such as J. C. R. Licklider, saw what would happen when we plugged computers into one another. We will live, Licklider told a gathering of world-class scientists, in a human-computer symbiosis, a coupling of symbol-manipulating networks that will be much greater than the sum of its parts.

As brilliant as those scientists were, many mocked his prophetic vision. Prisoners in their paradigms, they couldn't see past the bars of their cages.

Visionaries who see imaginary forests and foresters who tend real trees have different gifts.

Most of us miss the significant beginnings of New Things until long after they have passed. Beginnings are full of mystery and promise, darkness brooding on darker waters, while endings can be easier to spot. Toward the end of a trajectory, just when we have grown accustomed to "how things are," it will dawn on us that the trajectory is losing energy. We pause in a moment of reflection, realizing that a time of life, a way of being, a comfortable structure that had sustained us ... is not what it used to be. Not what it was.

Of course, every ending is also a beginning. Autumn leaves on the fast-flowing stream are interlaced with seeds. Nothing vanishes, nothing disappears. The universe is an engine of transformation that conserves everything, everything, even the light of dying stars as it streams toward black holes.

Horicon Marsh is one of the beautiful environments of the state of Wisconsin, a vast expanse of wetlands that stretches from horizon to horizon. My wife and I recently walked out into the marsh on a warm day, celebrating our anniversary.

It was late afternoon. The sun was low in the sky, sinking toward the flat expanse of still water. Our shadows lengthened in the tall grass and we paused, listening to the silence. The world around us was teeming with life. The edge of the light of the late afternoon seemed to be translucent, a boundary of our mutual journey.

In the spring and in the autumn, hundreds of thousands of migrating geese fill the sky above the marsh from horizon to horizon. They emerge from a vanishing point beyond the horizon into our short-term vision and fly overhead with cries growing louder and softer again as they disappear toward another vanishing point in the opposite direction.

Our children have all left home. Our decades of being a blended family have flown through the sky toward a vanishing point beyond the horizon.

One of our sons, nearly twenty years ago, yelped with regret after reading Stephen Levy's *Hackers*, a chronicle of the days at MIT when the seeds of Silicon Valley were sown.

"Oh, Dad!" he cried. "I was born too late! It's all over!"

And it was. The digital world had arrived for us through the narrow aperture of an Apple II, through which we crawled at 300 bps on ASCII Express toward BBS islands in the narrow dark seas of cyberspace. The world of hackers for which he grieved was fused with an exploration of inner space defined by mainframes that had already morphed.

The progressive death of each successive network has engendered another. Death and life embrace in a slow dance, cheek on cheek. The music ends. The music begins.

In response to the column "Digital Autumn," a friend in Australia wrote:

"You write of autumn, but I look around at Spring, which is well underway. The sun is higher in the sky, greens seem greener than a week ago, and nearby I can still enjoy the seldom occurrence of a whale that frolics in our harbor. Slowly the days lengthen and I seem to live in a world where our real communities are becoming more and more fragmented while digital communities are drawing us closer together. But how can this replace the ability to reach out and touch one another and look into another's eyes and share mutual moments in silence?"

Under the thin ice that forms on our hearts at the first frost, new spring life grows in the soil of irrevocable dissolution.

Autumn spring.

Life begins at edges, on the boundaries, at the interface where solitude is redeemed, transformed into community. When we touch one another and share a moment of silence, we feel the world fall away with a dying fall, we shift back into the rocking chairs of our souls, letting the inexplicable rhythms of life rock us to an unexpected beat.

Everything is ending, everything is beginning. The Internet is dissolving into the interstitial tissues of our symbiotic human/machine body/mind. The digital image of the universe filtering into our brains is a spiraling wheel of stars that fills the sky.

All of the symbol systems in the world, not just one religion but all religions, not just one civilization but all civilizations, are falling into this moment of transformation, into this black hole at the center of our galaxy. They emerge from a vanishing point beyond the horizon and disappear toward another vanishing point in the opposite direction. The body/mind conceives of a luminous cloud of new illusions, phantasms and imaginings that we will once again mistake for real gods and real demons. We can only pray that this digital pier on which we are standing will become a bridge across a galaxy teeming with life. We are like toddlers coming down the steps of our house for the first time. We are lesser lights of all those who have learned, who are learning to fly, our outward migration pouring out of the deep cave of our home planet at twilight. Filling the sky from horizon to horizon.

Our ancient identities cracking open, litter of broken shells feeding the fertile soil.

for Shirley, on our anniversary

A Miracle by Any Other Name

If any column is about "the human dimension of technology," it's this one, inasmuch as last week, my beloved youngest son Barnaby had more tubes in him, more drips dripping, more monitors flashing around him than a cyborg out of Terminator 3.

When I arrived at the ICU and saw, moving among the noisy machinery, his still-pink hand, swollen and slow, as it reached for my hand, I cried like a baby. In such moments the fragility, transitory nature, and absolute value of life, all life, is unmistakable.

My son was riding his motorcycle on Highway 101 in California when he came around a curve into stopped traffic. He hit the back of a pick-up truck and flew through the air. When paramedics arrived at the scene, there was no blood pressure and they pumped him full of fluids and kept him as stable as they could until a helicopter flew him to Modesto where they scanned the damage and decided that a torn aorta was the most critical injury.

He went into emergency surgery to repair the aorta. They gave fair warning that ignoring his badly broken leg might mean the loss of the leg, that bleeding from his liver had to wait, that staunching the blood flow to the spine during surgery might mean he wouldn't walk again. There is nothing to do when they read your rights but nod and sign off and get out of the way.

They repaired the aorta. The liver stopped bleeding. They operated on his shattered leg. They left alone his broken ribs and a crack in his upper back. They removed the ventilator and after a few days stopped the morphine drip. His vital signs are good. There's a long road ahead but it looks as if he'll make it.

Anyone who has been in an Intensive Care Unit lately knows that it looks like Ridley Scott designed it. Machines breathe, monitors regulate blood flow and drugs, cuffs flex and contract. It's like a scene out of *Bladerunner*, with robotic friends manufactured by canny engineers, friends that keep us alive.

Among the tubes and flashing lights is the reason the technology exists, the human soul in the machinery. Without my son's beating heart, which continues to beat, thank God, the high-tech devices would have no meaning.

The prognosis according to one of the docs is "fantastic." A torn aorta is fatal 85% of the time. With the other trauma, he said, there had been perhaps a 1% chance of survival.

My son can move his arms and legs and when he speaks it is obviously still my son with his characteristic genius for insight, understatement and humor. A devout Buddhist who has meditated for long hours at the Zen Center and Tassajara Monastery, he of all people can handle a view of a white wall, watching his mind and its shadows move.

We believe he will be OK and we are afraid to believe he will be OK. The depth and intensity of our own trauma, sourced by those telephone calls from hell, continues to linger.

Most of the doctors and nurses use words like "incredibly lucky" but some speak of a miracle and mean it. I hesitate to use that word lest those who lost loved ones wonder what happened when they could have used a miracle too. I do not pretend to understand how it all hangs together or makes sense. The older I get the more obvious it is that those who think they have a clue do not have a clue and those who know they do not have a clue have a shot at having a clue.

But in and of itself, that my son is alive and himself, that he will walk and talk and live, is a miracle by any name, whatever you want to call it.

Miracles come in many forms and during this hard time they sometimes came as felt realities, palpable touches of the spirit. When many people pray, express concern and love and are aligned in a single direction, their energy is amplified. When our consciousness is stripped of trivial concerns by the bone-deep clarity of a crisis, it enables us to focus with a laser-like intensity. When you feel those forces entering your awareness it feels like thermals during a hang-glide coming up from under. It feels like being lifted in a wave, like being a self-conscious node in a network aware of all the connections, knowing the pattern of the pattern of the web.

Our gratitude is impossible to express in such moments because it is absolute and words make everything relative. The choice of people to be there for us is sheer gift and grace and it is impossible to underestimate the impact of a kind word or a prayer. The extremity of our need may magnify the felt power of this unmerited benevolence but even in normal mundane everyday life compassion and generosity of spirit are the glue of the universe.

Anyone who believes the universe only works bottom up and not top down as well is missing some of the data. It begins and ends with consciousness as surely as a network map includes an image of the Big Picture as well as nodes feeling each other out, knitting themselves together from all sides. When we extend ourselves toward each other's needs we make a connection, becoming something more for a moment but in fact becoming only what we have always been, a singular being not always fully aware of itself in all its particulars, alive in a universe more like thought than stuff or maybe thought and stuff at the same time. As I said, I really haven't got a clue, just an inkling, an inkling made as bold as the brush stroke of a Zen master on an empty canvas by a moment of transparent clarity and utter terror.

Talking to Ourselves

Once upon a time in the sixties, I published a short story in *Analog Science Fiction* about a man who invented a virtual reality machine and let a carnival owner try it out. The carnival owner was so hypnotized by the fantasy world and its contrast with the grim realities of his life that he never wanted to leave the machine.

Twenty-plus years later I was watching the *Twilight Zone* and was startled when a short episode consisted of my story. There was the machine, the person who wanted to stay in it, everything. Obviously, I thought, someone had read the story and used it as the basis for the television program—without, however, paying royalties.

I contacted the producers and received a telephone call from Harlan Ellison. I was star struck because I had been reading and admiring his science fiction for years, so I was predisposed to believe what he told me.

The story had evolved through parallel evolution, Harlan insisted. He sent me a stack of scripts that showed how the story had changed from version to version and arrived at a similar form by a different route. I accepted his explanation.

Recently I reviewed some stories written in my forties to see if I could revise them. One called Learning Curve used the device of a time machine to show how a middle-aged man looked back on mistakes he had made growing up, mostly due to decisions he made about how to interpret what was happening. Those decisions determined his identity, and identity is destiny—who we think we are is how we choose to express our lives—so the middle-aged man went back to critical moments in his history to try to teach his younger self a better way, to give himself the benefit of his wisdom.

Each time he tried to teach himself a lesson, however, it was beyond the grasp of the younger person. At the end, when he ventured to the recent past and saw himself trudging up the walk to his home, his spirit deflated by a recent divorce, he realized he had nothing else to say; all he could do was step out of the shrubs, tell himself that he understood, and hold his younger self while it cried.

Once he learned to be compassionate toward himself, he stopped looking backward and could move forward again.

That story had promise, I thought, because of its emotional truth. But I am not going to revise it because, in the years since, a movie starring Bruce Willis called *The Kid* told a similar story. People would think I had stolen a plot device that was already trite when I used it.

The writers of the script arrived at a similar plot by parallel evolution, I imagine, and if I could see the revisions of the script before it took final form, it would probably resemble what Ellison had sent.

A decade later, I relate the story as much to mentoring as to introspection. When we mentor someone, particularly when we are in our fifties and they are in their twenties and thirties, we are talking not only to them but also to our younger selves, but this time with compassion. Some mistakes are inevitable, we know, and we can't learn what we need to know without the detours that turn out to be the most direct route to maturity.

Several of our children were with us last month for the holidays and it was a wonderful time. Everybody is an adult now and we relate to one another less as parents and children than as older and younger adults. My youngest son whose motorcycle accident I described in a former column, writing in the waiting room of the ICU after we knew he would live, he would live! was one of our visitors. He is healing and we are grateful.

My son and I stopped for a latte late one afternoon. Through the coffeeshop window on the Milwaukee River, we could see the woods which had not yet filled with snow and the cold river. We watched the trees thicken and dissolve in the dusk as we talked about the transitoriness of all things, his deep and terrible insight into the nature of appearances and what we call realities. We tried to make our lattes last, stretch the canvas of time over the frame of the lengthening shadows. We spoke the most real things we know. But at last the twilight knitted the trees into an inscrutable darkness and we had to go.

That conversation is a self-referential image, it is what it was about. It is a transparent stained glass window illuminating movement with illusory images of fixity.

Was there any way for either of us to learn what we know by an easier route? Was there any way for me to relieve the ones I love of the necessity of traveling their own roads? I guess not. As my forever-unpublished story pointed out, the only road is the one we walk. The knowledge of the impermanence of all things creates in us if we are paying attention a compassionate heart which is willing to listen and understand and at the end embrace and feel both love and the hard edges of boundaries defining our individual destinies.

There is no greater joy than loving and mentoring when we can, but I know now that whenever I mentor another, I am really healing my younger self who longed at critical moments for someone older, more knowing, more loving, to show up and just be there. The grace of my life has been the realization that whenever I really needed that, someone did. Their faces are in the lighted hallways of my memories, portraits of my own personal saints, the ones that matter most. Nor does it diminish their contribution or value to know that they too reached their mature selves by a process of evolution parallel to my own and that they too were talking not only but also to themselves.

Coming of Age

The number isn't important, friends have been saying when I talk about turning sixty. Some say, age is only a state of mind. Some say, you're as young as you feel. Some say, age doesn't matter.

And some say, why, you look great! which unfortunately confirms that there really are three stages of life: youth, middle-age, and you look great!

Well, my well-intended friends, I am here to tell you that age does matter. In some ways, it matters a LOT.

When older people and younger people talk, they look at each other differently. Younger people have a shorter gaze.

I was taught the meaning of a long gaze by a high school teacher, Miss McCutcheon, who gave me her long teacher's gaze during an English class. I felt like a butterfly, pinned and wriggling. When another student asked what she was seeing, she said simply, "Some day someone is really going to love that boy."

I couldn't handle that. I was fifteen, fat and self-conscious and confused, and I squirmed, turned red and snapped something back … but have never forgotten what she said. At a time when love seemed beyond my reach, her insight was deeper than mine, living as I did half-blind and half-crazy in an adolescent storm of rain hail and thunder.

Coming up to sixty, we see other people, especially younger ones, more often with that long look. We see who they are and who they can become if they only attend to the better angels of their natures.

Sometimes there are moments during such conversations when it feels as if the years fall away and transparencies of other conversations, ones that happened years ago, meld with the one I am having now. Memories control the present moment, capturing it with a force field of longing and grief before the experience becomes transparent to its underlying dynamics, the irrevocability of my own past juxtaposed with seemingly innumerable futures for the one to whom I am speaking, branching like blossoms of forbidden opportunity.

Then the regret fades, replaced by encompassing acceptance of the only life we have to live.

We may not know how to say what we know in such a moment, but we do know and we know that we know.

We are no longer innocent, coming to sixty. We know what evils can befall us. We know as Robert Frost said that there are finalities besides the grave. We know ourselves, sometimes too well. We remember too many people we have loved and held as they died. Somehow the degree to which we have lived with passion and gusto informs our awareness of death as well as our love of life.

Two recent movies, *Lost In Translation* and *Eternal Sunshine of the Spotless Mind*, capture magnificently the poignancy of moments of love and loss, showing connections deep bone-in-the-socket solid for only a moment before the whirlwinds of our lives take us again in different directions.

I love good films the way I loved good books as a child. Coming up to sixty, I accept that being a latchkey kid and losing my parents early gave me a particular destiny. William Gibson, the cyberpunk writer, notes at his web site that many writers share that kind of loss or other early childhood trauma. We find solace in the world of imagination and images, building meaning from the tools at hand. Books and films provide points of reference for sharing insights, giving us a common language.

In another great film, *My Dinner with Andre*, Wallace Shawn and Andre Gregory say that a moment of genuine connection with another person heightens our awareness of being alone, too, and to accept that we're alone is to accept death, because somehow when you're alone you're alone with death.

That's the implicit affirmation when the protagonist of *Eternal Sunshine* says OK. OK. to a doomed trajectory of romantic love. It is also the moment of genius at the end of *Lost in Translation* when the director/writer makes the final words whispered by Bill Murray in the ear of his young friend impossible to hear.

It doesn't matter what he said, and it doesn't matter what I say either. We always fail to articulate what we nevertheless unceasingly try to say, the deepest truths we know, which can only be suggested like the moment of waking from a dream more real than the sunlight streaming through the window, when we know we will never remember the dream exactly but nevertheless have another day, another day, another day in which to pursue it.

Coming to sixty does make a difference. It is clear that what we mistook for achievement is empty air, unless it made a real contribution, unless it made a difference, it is clear that mostly self-serving efforts deliver as much satisfaction as drinking from a dribble glass.

Still, we are left with questions, not answers.

Which were the moments of genuine self-transcendence in which I was called to be more than I thought I was and somehow fulfilled the promise? Which did I miss? What is possible in the time left, as eyesight fades but the sharper-eyed inner gaze of an ancient mariner discerns with greater clarity what matters most?

If we are fortunate, the choices we make now, coming to sixty, were determined many years ago when earlier decisions built the karma of our destiny. We all fail, and we all succeed. There is nothing now but the sudden unexpected opportunity, nothing but being ready. There is nothing to hold back, no energy to save for another day.

I know that I am alone with life and death. Even in moments of the deepest communion, I can feel the world turn and the spiraling universe bend away from my embrace. Moments of dizzying lucidity, seeing the anchor of the life given and the life received for what it is, counterweight or ballast, nothing amassed.

Made in the USA
Monee, IL
21 November 2024